APPLIED CHAOS THEORY

A PARADIGM FOR COMPLEXITY

APPLIED
CHAOS
THEORY
A PARADIGM FOR COMPLEXITY

A. B. ÇAMBEL

George Washington University
Washington D. C.

ACADEMIC PRESS, INC.
Harcourt Brace & Company, Publishers

Boston   San Diego   New York
London   Sydney   Tokyo   Toronto

Copyright © 1993 by Academic Press, Inc.

ACADEMIC PRESS, INC.
1250 Sixth Avenue, San Diego, CA 92101-4311

United Kingdom edition published by
ACADEMIC PRESS LIMITED
24-28 Oval Road, London NW1 7DX

Library of Congress Cataloging-in-Publication Data

Çambel, Ali Bulent, 1923.
    Applied chaos theory : a paradigm for complexity / A. B. Çambel.
        p.   cm.
    Includes bibliographical references and index.
    ISBN 0-12-155940-8
    1. Science—Philosophy.   2. Complexity (Philosophy)   3. Chaotic behavior in systems.
I. Title.
    Q172.5.C45C36   1993
003'.7—dc20   92-27983
                                                                                                    CIP

Printed in the United States of America
92 93 94 95 96   BC   9 8 7 6 5 4 3 2

I dedicate this book to
the memory of my parents, who taught me structure and chance,
and
the future of my grandchildren—may their chaos have high dimensions.

TABLE OF CONTENTS

PREFACE

Complexity is ubiquitous. It is in nature as well as in artifice. It occurs in large and in small systems. It can be tangible or intangible. To be aware of the existence of complexity can be like feeling a "presence" that virtually defies description.

As yet, there is no agreed-upon explicit definition of complexity, although there are various operational descriptions. How much we frail human beings are meant to understand about complexity is nebulous. The task is both awesome and formidable. Indeed, it may be far too presumptuous to insist that we can explain everything. Perhaps all one should hope for is embodied within a statement by the guru of the arts, sciences, humanities, and governance, the Science Advisor to Presidents Kennedy and Johnson, and President Emeritus of M. I. T. Jerome B. Wiesner:

> Some problems are just too complicated for rational, logical solutions. They admit insights, not answers.

It is in this spirit that I have written this book. I have attempted to explore the potential applications of chaos theory to reveal "insights" into the structure and the dynamics of complex systems. Like any other methodology, chaos theory has its limitations, and is still in its formative stages. Accordingly, I have stepped outside nonlinear dynamics per se, and have included nonequilibrium thermodynamics, information theory, and fractal geometry. This is natural because complexity transcends the boundaries of traditional disciplines.

I make no claim to originality. I have attempted to emphasize applications and have shied away from both erudite mathematical discussions and details of the mathematical elegance. There are outstanding books that focus on the mathematical process. I subscribe to the school of thought that a person who is familiar with a subject can learn it better. Accordingly, I have been more concerned with the potential applications, meanings, and limitations of equations. I consider this important in dealing with an interdisciplinary subject. By definition, anything that is applied and interdisciplinary loses its purity and becomes contaminated. This should not be deplored, because complexity itself is not pure; it is replete with paradoxes.

I have presented the material at an introductory level so that a broader audience can take advantage of it. However, I have provided extensive references to assist the peripatetic reader in locating the original sources. I do not insist that chaos theory is the only vehicle through which light is shed on complexity. There are other approaches, such as catastrophe theory, cellular automata, or synergetics, that might be equally efficacious. Chaos theory, together with its compatriots that I have included, lends itself to the study of complexity because nonequilibrium, nonlinearity, and unpredictability are major characteristics of complex systems.

Professor John Guckenheimer[1] of Cornell University, who has contributed so manifestly to the modern formulation of nonlinear dynamics, has rightfully questioned whether chaos is a science and has articulated its strengths and limitations much better than I can. On the other hand, there is much truth in the 1986 statement by Sir James Lighthill,[2] who occupied the Lucasian chair in mathematics at Cambridge University that Isaac Newton once held. Speaking before the Royal Society on predictability, he stated:

> . . . the enthusiasm of our forebears for the marvelous achievements of Newtonian mechanics led them to make generalizations in this area of

---

[1] Guckenheimer, J. (1991). "Chaos: Science or Non-science?" *Nonlinear Science Today*, 1 (2), 6–8.

[2] Lighthill, J. (1986). "The Recently Recognized Failure of Predictability in Newtonian Dynamics," *Proc. Royal Soc. of London*, **A407**, 35–50.

predictability which, indeed, we may have generally tended to believe before 1960, but which we now recognize were false.

The jury is still out as to whether or not chaos theory will provide solutions to complex systems. It is conceivable that a new mathematics will be developed to explain complexity, as has occurred in the past to describe other facades of science. Examples include the formulation of calculus by Newton and Leibniz, the development of vector analysis by Josiah Willard Gibbs, the discovery of Fermi–Dirac statistics, and formulation of Feynman diagram techniques.

Chaos theory is best practiced when supported by computers and appropriate software. The former can be expensive, and software development is time-consuming. But willing people should not be deprived of opportunities to probe complexity. Accordingly, I searched for modestly priced software programs written for affordable personal computers. The software programs referred to in this volume are for IBM™ or compatible PCs. However, modestly priced programs for the Macintosh™ are also available. Programs may also be found in the public domain. It was in the same spirit that I did not include color plates, which require elaborate equipment and would have raised the cost of this volume. I hope that my penury will benefit solo researchers, small groups of scholars, modest institutions, and persons in one field or another who dare to apply the new paradigm to their traditional vocations.

I believe that society would prosper if we learned together, unshackled from disciplinary ethnicity. To be able to do that we need a common idiolect. Chaos theory is one such mode of communication. An interdisciplinary approach to chaos theory bolstered by nonequilibrium thermodynamics, information theory, and fractal geometry transcends disciplinary lines and has been found useful in multifarious areas. Accordingly, this book is not aimed at any particular disciplinary audience, and no specific prerequisites are necessary to understand its contents. The examples I have cited emanate from different fields that we are close to as individuals, rather than being in the domain of specialists. Also, I have included historical highlights because I believe that a literate person must be appreciative of the traditions that gave momentum to the development of the particular body of knowledge. It will be evident from these remarks that the presentation is not parochial. Such an attitude is necessary because the tentacles of complexity are many, and they entwine extensively. It is my hope that the book will encourage interdisciplinary courses and/or seminars. Also, I hope that the solo scholar will find the book useful as a self-help medium. Finally,

I hope that the peripatetic reader will curl up in front of a fireplace, or stretch out in the seat of a nonstop flight, and peruse these pages.

Most of the material covered here was developed for a beginning, one semester interdisciplinary course at the George Washington University; the course has been attended by students from some of the universities that are members of the Consortium of Universities of the Washington metropolitan area. A wide variety of disciplines have been represented by the attendees, who have been undergraduates as well as postdoctorates, and I might note that their ages ranged from 17 to 70. Some parts of this text were presented at a United Nations seminar, Wayne State University, the Federal Institute of Technology in Zürich, the University of Florence, the University of Mannheim, the Institute for Advanced Study in Berlin, the Smithsonian Institution's Campus on the Mall, the National Institutes of Health, the Veterans Administration, the Washington Evolutionary Sciences Society, and the Cardiology Update course in Northern Virginia. The reactions of these diverse groups broadened my intellectual horizon in numerous ways.

A. B. Çambel
McLean, Virginia

PREFACE TO THE SECOND PRINTING

This second printing made it possible to correct several typographical errors. I am indebted to J. Long, C. T. May, M. Shlesinger, C. Stricklin, and R. C. Warder for pointing these out. It should also be noted that the software programs of W. M. Schaffer and his associates, which are cited in the references, can now be obtained from Campus Technology, 751 Miller Drive, SE, P.O. Box 2909, Leesburg, VA, 22075; 1-800-543-8188.

A. B. Çambel
September 1993

ACKNOWLEDGMENTS

The late dean of broadcast and television journalism, Edward R. Murrow, had a series wherein he would interview significant persons. One such person was the great singer Marian Anderson. Ms. Anderson would answer all of Mr. Murrow's questions with the word "We . . . ." Finally Mr. Murrow asked why she did so, particularly because she was so well known as a soloist. Ms. Anderson replied that there was very little that anybody could do alone, and that even as a soloist she sang the songs of composers, had the support of her accompanist, relied on the hall acousticians, could not do without the stage hands, and could not excel without the enthusiasm of her audiences.

So it is with this book. I could not have written it without the help and encouragement of others. In citing them, I shall start with my family. My wife, Marion, provided inspiration—by sharing her expertise in landscape design—about order coming out of chaos, and about the prevalence of fractals in nature. She read the first and final versions of the manuscript and made invaluable editorial improvements to my polyglot syntax. She was understanding throughout when I was frustrated by technical conflicts among my computer, scanner, and printer while trying to compose the text. She was encouraging at all times although we have been married for almost half a century. My son Metin, our eldest, tweaked my computer to make it perform according to my expectations and was available at his end of the telephone line to advise me about short-cuts. My daughter Emel provided serenity in her Vermont farm home amidst her lively family and her horses when I needed to get away, far from the madding crowd. My daughter Leyla went through the entire manuscript, made corrections, and gave me gentle tutorials on how to handle manuscript preparation; she also performed as my volunteer agent in obtaining copyright permissions and attending to myriad other matters. She was a bulwark of strength throughout and a source of solace. Her deep friendship and tender loving care were boundless. My daughter Sarah, our youngest, always showed curiosity about the progress of the project, and nurtured my determination. She shared her expertise about desktop publishing and provided imaginative advice. Her youthful sagacity, unscarred and unafraid, was a potent infusion of spiritual strength. She kidnapped me to vacation retreats, took me to plays, concerts, and to parties, and cheered me on as only a close friend knows how to do so tactfully. Although not immediate members of my family, three persons have kept me alive physically without loss of dignity: namely, F. H. Andersen, M.D., my cardiologist; P. M. Borges, M.D., my oncologist; and H. Lowe, who held me in her healing powers. I am grateful to all of them.

This book would not have been conceived without the stimulus of three individuals: B. Fritsch, of the Swiss Federal Institute of Technology in Zürich, helped me comprehend that physical and economic sciences share similar frustrations and aspirations. F. A. Koomanoff, of the U. S. Department of Energy, made me aware of the impacts of uncertainty on matters that affect the commonweal. He also gave generously of his time to review different chapters of the manuscript. J. Sommer, of the Department of Housing and Urban Development, prodded me gently to get involved in self-organization, supplied me with resource material from his own fields of interests and was supportive in many ways. I owe them much for these awakenings.

I am indebted to three department chairpersons at George Washington University, R. R. Fox, C. M. Gilmore, and A. M. Kiper, for their encouragement and their defense of nontraditional research. I am grateful to C. Long for introducing me to the ALADIN System of the Marvin Gelman Library, which made it possible to search for publications by modem in the middle of the night from the sanctuary of my study at home. I am grateful to Q. A. Amiryar for always being able to obtain publications I needed. I thank G. W. McIntyre and R. E. MacLean for administrative support in dealing with grants. I am indebted to Z. Coles, our office manager, for her numerous thoughtful courtesies facilitating my work. Three graduate students, A. Ghassam, A. Parvin, and W. Pedrosa, helped locate references.

A number of friends and colleagues read all or parts of the manuscript and provided invaluable advice and criticism. J. Feir of George Washington University reviewed several chapters with his usual meticulous care, and he was always helpful in clarifying subtle mathematical issues in front of the chalkboard or on the back of paper napkins during lunch. L. Hodges of Georgia Institute of Technology clarified my misconceptions about scaling and fractals, and helped me restructure several chapters. R. Goulard of George Washington University, C. Hauer of the National Science Foundation, and A. K. Oppenheim of the University of California (Berkeley) insisted that I recognize better intellectual giants who contributed to complexity and chaos theory indirectly but indisputably. I am indebted to S. S. Penner of the University of California (San Diego), who advised me on several chapters, pointed out references I had overlooked, and clarified the thrust of the project. I thank F. W. Summers of St. Joseph's Hospital, Orange, California, a person of rare breadth and depth of knowledge, for exchanging views on all sorts of complex matters while we were vacationing on Martha's Vineyard. He read the entire manuscript, enticed his colleagues to read it, and pointed out parts that needed strengthening. I value his friendship and camaraderie. J. Summers of Summers and Associates, Port Washington, Long Island, not only critiqued selected chapters, but also helped me improve my skills in scanning graphics, and was generous in sharing his expertise about modern desktop publishing techniques. In the process we also developed a mutually supportive and liberating friendship which I treasure.

I extend my sincere thanks to several persons who went out of their way to locate illustrations. They are: B. Bastek of the Austrian Embassy, M. Boyd of the University of Texas, S. Beddow of M.I.T., T. A. Boden of the Oak Ridge National Laboratory, D. Fell of Derek Fell's Horticultural Picture Library, H. Kanamori of Caltech, N. O'Brien, J. Sim of Bettman Archive, A. Shurygin of the Russian Embassy, and F. Zeitlhofer of the Austrian Cultural Institute.

I value stimulating discussions with T. Bernold of Zürich; J. Chandler of the National Institutes of Health; B. Dendrou of ZEI, Inc.; J. Eftis of George Washington University; K. S. Madden of Children's Hospital; J. E. Mock of the Department of Energy; S. A. Schuh of the University of Tampa; and S. Stecco of the University of Florence. At times these stimulating dialogues raised questions no author likes to face, but which nevertheless are crucial to clarifying the presentation of the text material.

Lee A. DuBridge, that thoughtful statesman of science, former presidential science advisor and President Emeritus of Caltech, remarked that the best friend a teacher has is a good student. I have been blessed with many. While I was working on this manuscript, D. W. Barns, S. A. Basile, B. D'Amico, S. Durbano, C. R. Fulper, S. R. Kasputis, S. Lesher, B. Loew, P. Kurzyna, F. Menandro, S. Moore, A. Oni, J. Reeke, M. Pena-Taveras, P. Vitale, and W. F. Zeller, III never ceased to stimulate me with their questions, and at times they confirmed with their computational expertise the behavior of complex systems that I intuitively asserted.

A number of organizations directly and/or indirectly helped me to learn about complexity. First of all, I must thank George Washington University for being so cooperative to a faculty member who takes early retirement to concentrate on a subject he wants to learn. Other organizations were most supportive and include the Allied Bendix Aerospace Technology Center, Sigma Xi, the Scientific Research Society, the Department of Energy, the Sandia National Laboratories, the Liberty Fund, the Gottlieb Daimler and Karl Benz Foundation, the Deutsche Forschungsgemeinschaft, the Swiss National Foundation for Energy Research, the Mannheimer Forschungsstelle, and the United Nations Organization.

In the final analysis, research and writing should be shared with others, and this can happen only if the work is published. Of course, no manuscript is transformed into a book by itself. Thus it is with great pleasure that I acknowledge the many imaginative contributions of my editor and publisher, David F. Pallai of Academic Press. He harmonized the needs of his firm with the aspirations of this author and the expectations of the reader without ever compromising his integrity. The quality of the book has benefited distinctly from his advice, guidance, and supervision. The late Maxwell Perkins, the dean of editors, could not have been more influential. The Academic Press, Inc., staff was supportive throughout and, in particular, I must thank Brian Miller who was in charge of the production of the volume. He brought his high standards to bear on the diverse aspects of the task, his competence overcame many an obstacle, and he was always ready to accommodate the author's needs. I would also like to thank Pascha Gerlinger, Karen Pratt, and Christina Wipf of Academic Press, and the copyeditor Kristen Cassereau.

I suspect that in spite of all the advice and assistance I have received there are errors of omission and commission. Only I should be blamed for these. I invite readers to bring their comments and advice to my attention.

A. B. Çambel

CHAPTER **1**

LIVING WITH COMPLEXITY

INTRODUCTION

We live in a world that is so extraordinarily diverse in composition, form, and function that any standard classification is impossible. Whether we face our daily obligations with alacrity or muddle through the day, we cannot help but be aware of the complex nature of all that surrounds us and the uncertainties that we face. Even within our immediate personal universe, we are not at complete liberty to do as we please, because we are inextricably linked to our natural environment, the social and professional institutions with which we are associated, and even the myriad technological devices that we use. Thus a little push here, a little pull there, and major events can occur, some for the good, and some not so pleasant. In other words, cause and effect are not proportional. A small effect can have significant consequences; conversely, a major effort might yield very little. Mathematicians call such events *nonlinear*. Usually complexity involves nonlinearity. Unfortunately, there are no general explicit solutions to nonlinear differential equations, and each case must be treated on its own merits. It follows that formulating generally applicable models for real-life situations is not feasible.

Another aspect of complexity is that it involves chance. It is not necessary to buy a lottery ticket to be confronted by a chance event. We are bedeviled by chance and probability even when we do not seek them. We are accustomed to experiencing unexpected joys and/or misfortunes. Intrinsic to complexity is the paradox between the orderly, deterministic nature of scientific laws, and the chance events that occur all around us and in all aspects of our experiences. Scientific laws are really not as deterministic and categorical as they are considered to be. Actually, there is always some fine print specifying the idealizing assumptions made. How then can an event obey the orderly laws of physics as well as the laws of chance? How is it possible that there can be random components in deterministic events? How can we deal with their coexistence?

What this means to us as individuals is that in the dynamics of our daily lives uncertainty is quite normal. This does not mean that it is futile to plan. On the contrary: Once we accept that some element of chance is unavoidable we must develop improved ways of forecasting. Complexity is omnipresent, but dealing with it is not simple. The principle of bounded rationality enunciated by Herbert Simon,[1] the Nobel laureate in economic science, states:

> The capacity of the human mind for formulating and solving complex problems is very small compared with the size of the problems whose solution is required for objectively rational behavior in the real world . . . .

CHARACTERISTICS OF COMPLEXITY

We have no agreed-upon definition of complexity, because it manifests itself in so many different ways. Accordingly, operational descriptions of complexity are helpful. How can one define in a few words a concept that appears in so many different ways? There is complexity in nature as a whole, in each of its species, in the myriad of devices and process devised by man, and in the social institutions that are meant to be helpful. Of course, beliefs and ideas, too, are complex.

In spite of this variety there are certain basic characteristics that must be considered. These are: (a) purpose and function; (b) size and configuration; (c) structure, including composition and makeup; and (d) the type of dynamics. I shall call the first two the "static complexity," the third the "embedded complexity," and the last the "dynamic complexity."

For total complexity these factors should coexist; however, this need not be. For example, consider the World Trade Center. During weekends all is quiet, and the dynamics are steady and relatively minor except for maintenance functions. However, during working hours there is the hustle and bustle of persons working and visiting in the building, and the communications within the building and with the outside world, as well as the restaurants and shops, spring into action.

The dynamics of a complex system may vary, and typically we encounter this when during travel we cross different time zones and our normal biological rhythms lag or advance. Depending on the circumstances, dynamic stability may be steady, transient, or chaotic.

Structural complexity in itself is an important characteristic. It may be argued that an elephant weighing one ton is more complex than a one-ton rock. As geologists will tell us, not all one-ton rocks are equally complex. In the biological domain, were the extinct trilobites who seem to have had complex structures and functions less complex than homo sapiens? Was the 1937 Mark I computer built at Harvard University, which weighed 5 tons, had 500 miles of wiring and 3,304 electromechanical relays, less complex than a present-day, state-of-the-art desktop computer that can do so much more and at far greater speed? Is a drug like AZT more complex than penicillin or aspirin? Is it its chemical structure or its kinetics in the body that makes the difference? It follows that the complexity of a system must be considered in the light of the surroundings in which the system finds itself.

Some examples of complex problems that we are likely to encounter as we go about our business are traffic flows, weather changes, population dynamics, organizational behavior, shifts in public opinion, urban development and decay, cardiological arrhythmias, epidemics, the operation of the communications and computer technologies on which we rely, the combustion processes in our automobiles, cell differentiation, immunology, decision making, the fracture of structures, and turbulence. Quite obviously these are all rather different types of events. This is why complexity cannot be neatly packaged into a standard container, but must be dealt with heuristically.

The following statements can be made:

1. Complexity can occur in natural and man-made systems, as well as in social structures.
2. Complex dynamical systems may be very large or very small; indeed, in some complex systems, large and small components live cooperatively.
3. The physical shape may be regular or irregular.

4. As a rule the larger the number of the parts of the system, the more likely it is for complexity to occur.
5. Complexity can occur in energy-conserving systems, as well as in energy-dissipating systems.
6. The system is neither completely deterministic nor completely random, and exhibits both characteristics.
7. The causes and effects of the events that the system experiences are not proportional.
8. The different parts of complex systems are linked and affect one another in a synergistic manner.
9. There is positive or negative feedback.
10. The level of complexity depends on the character of the system, its environment, and the nature of the interactions between them.
11. Complex systems are open in the sense that they can exchange material, energy, and information with their surroundings.
12. Complex systems tend to undergo irreversible processes.
13. Complex systems are dynamic and not in equilibrium; they are like a journey, not a destination, and they may pursue a moving target.
14. Many complex systems are not well behaved and frequently undergo sudden changes that suggest that the functional relations that represent them are not differentiable.
15. Paradoxes exist, such as fast and slow events, regular as well as irregular forms, and organic and inorganic bodies in cohabitation.

It is not difficult to understand that something this varied is difficult to describe and even more difficult to explain. Before I start discussing the ways of dealing with complexity, I must review some terms that are taken for granted but often are misunderstood.

Determinism, Randomness, and Uncertainty

According to Pagels,[2] complex systems fall someplace along a spectrum extending from perfect order to complete randomness. In reality, the existence of either extreme condition is difficult to achieve. Even simple systems tend not to be completely deterministic, and it is also questionable whether perfect randomness can be achieved. It is quite obvious that there can be no rational way of reaching a well-reasoned decision about a completely random event. On the other hand, decision making is preempted in a completely deterministic situation because we would be dealing with a foregone conclusion. There would be no surprises; our behavior in making plans, purchasing insurance policies, or indulging in the games of Lady Luck would be entirely different if either perfect determinism or complete randomness were to exist.

Randomness is in sharp contrast to determinism. By determinism we mean that an event is caused by certain conditions that cannot possibly lead to any other outcome. For a dynamic situation this means that given the initial conditions the trajectories can be calculated with reasonable precision. Stated differently, the data describing deterministic phenomena can be calculated with the use of explicit mathematical formulas. Throughout this book we shall be concerned with deterministic dynamical systems. These follow a rule that prescribes how the state of the system changes with time from one state to another. When the representative state points

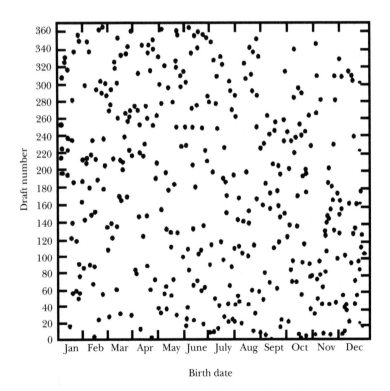

Figure 1. Scatter graph for 1970 military draft lottery. (From *FOR ALL PRACTICAL PURPOSES* by COMAP. Copyright (©) 1988 by COMAP, Inc. Reprinted with permission of W. H. Freeman and Company.)

are connected, one obtains the trajectory of the system. This trajectory cannot cross itself.

In a truly random process there would be no meaningful time series—there would be a scatter graph. An example of random behavior in the 1970 military draft lottery is shown in Fig. 1 from Garfunkel and Steen.[3]

Although the numbers drawn appear widely random, this evidently was not truly the case, because the capsules had not been adequately mixed. Hence draft registrants born towards the end of the year were more likely to have their numbers pulled. The matter was corrected in the following year. See Fig. 2 for the year 1971. While even this is not perfectly random, it is an improvement.

For a system to be truly random, there must be no causal relation between an observation at some present time, t_n, and a past observation at an incremental time, t_{n-1}, or an observation at an incremental future time, t_{n+1}. Because a set of numbers appears to be random does not mean that it is. Statistical analyses must be undertaken to establish the nature of the data. For example, it has been shown by P. Diaconis and D. Bayer that in card games a deck must be shuffled seven times before the odds that a card may be in any position are the same. In turn, two decks must be shuffled nine times for the odds to be evenly distributed (Kolata[4]). The reader who wishes to learn about random coin tossing should read the stimulating article by Professor Joseph Ford[5] of the Georgia Institute of Technology. The

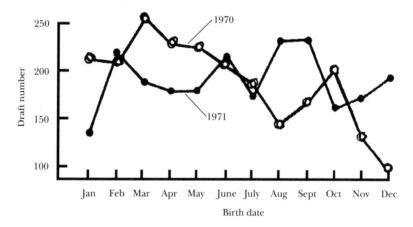

Figure 2. Median draft numbers by month. (From *FOR ALL PRACTICAL PURPOSES* by COMAP. Copyright (©) 1988 by COMAP, Inc. Reprinted with permission of W. H. Freeman and Company.)

problem is a fundamental one, namely the questionableness of having true randomness. What is called random is in reality pseudo-random. Even in the case of computerized random number generators, one can only expect pseudo-randomness, because the computer uses a prescribed pattern, which can be easily figured out.

LAPLACE'S DEMON

In the early days matters like complexity were debated by natural philosophers who were at once philosopher and scientist. This all changed with Sir Isaac Newton (1642–1727), whose contributions include differential calculus and the laws of classical mechanics. These gave renewed impetus to determinism, which traces back to Socrates (470–399 B.C.). Newtonian physics made it possible to calculate and thus determine the dynamics of objects by straightforward equations. Marquis Pierre Simon de Laplace (1749–1827) promulgated Newton's enunciations with his own contributions. This famous statement of Laplace (Kline[6]) had a particularly strong influence on setting the course of science:

> We may regard the present state of the universe as the effect of its past and the cause of its future. An intellect which at any given moment knew all of the forces that animate nature and the mutual positions of the beings that compose it, if this intellect were vast enough to submit the data to analysis, could condense into a single formula the movement of the greatest bodies of the universe and that of the lightest atom; for such an intellect nothing could be uncertain and the future just like the past would be present before its eyes.

This kind of deterministic belief was attractive not only to scientists, but to other sectors of society. After all, one could expound and dictate categorically, invoking the authority of the laws of science! Whereas Newton had tried to relate the science of the day to the Creator, Laplace had no such interests,[7] but rather wanted to prove that, like clockwork, the universe functions rationally according to the laws of mechanics. In Laplace's deterministic world there would be no uncertainty, no chance, no choice, no freedom, and no free will. Everything would be predetermined. We know from personal experience that this cannot be. How often have unexpected minor events changed our so carefully laid plans?

From the scientific viewpoint, strict determinism must be ruled out because measurements are affected by the presence of the observer. Even the so-called noninvasive measurements affect the system at least microscopically. We also know that the number of particles constituting any system is horrendously large, about 2.7×10^{19} particles per cubic centimeter, so that their coordinates and momenta cannot be specified except statistically. Even on the scale of the world's population, namely about 5.3×10^{9}, a much smaller number, we cannot tell the whereabouts, nor the activities of, individuals. There is always going to be some uncertainty.

It should not be inferred that uncertainty and randomness are synonymous. Quite the contrary is true. In essence uncertainty is a manifestation of information, or the lack thereof. No undue randomness may be present, but we may still be uncertain in evaluating a situation. Of course, if randomness is involved, the uncertainty increases. Uncertainty also tends to increase as the system under consideration becomes more intricate. The basis of randomness lies in probability theory, while uncertainty is related to information theory.

NONLINEARITY

Rarely do dynamic events follow a straight line for an extended period of time; eventually they exhibit nonlinearity. This occurs in different ways, and to provide some formal organizational mechanism it is customary to speak of *oscillators,* such as pendulums. To be more specific, one differentiates among different types of model oscillators such as the undamped, unforced, linear oscillator, which is the simplest, and the damped, forced, nonlinear oscillator, which is the most complicated. I shall not elaborate further on these, because I want to emphasize a different outlook. The interested reader will find an excellent descriptive comparison of the different types of oscillators in the fine volume by Thompson and Stewart.[8]

Complex problems invariably involve nonlinearity. Nonlinear events may be regular or irregular. Figure 3 shows a regular nonlinear trace depicting a sine wave.

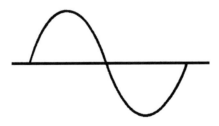

Figure 3. Sine wave.

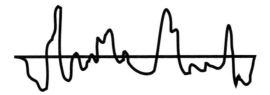

Figure 4. Representative trace depicting a sound wave.

In contrast, Fig. 4 shows a nonlinear curve that is irregular.

The above figure shows how complex even one note emanating from a single musical instrument can appear. If something that simple can look so convoluted, it is easy to comprehend why so many real-life situations appear so complex. One can generate all sorts of complex traces with a sound generator.

Unfortunately, there are no explicit general solutions to nonlinear mathematical problems. In the past, there was a tendency to deal with nonlinearity by considering such problems as aberrations and ignoring them. With increasing populations, dwindling resources, and rising expectations, we can no longer indulge in this cavalier attitude. We have to face the real-life problems that confront us and learn to deal with them. This is where chaos theory comes in. Among its many applications is its ability to provide insights to nonlinear phenomena that involve random aspects.

Consider the discrete logistic equation that we use extensively in Chapter 6. One common form of this equation follows:

$$x_{n+1} = ax_n (1 - x_n) \tag{1}$$

Here $\{x\}$ is a time dependent variable, and $\{a\}$ is a parameter that influences the degree of nonlinearity in this equation. Actually it tells us about the rate at which increases (or decreases) take place. The subscripts denote generations or increments of time. Thus $\{n + 1\}$ refers to the generation that follows the nth generation (e.g., child), $\{n + 2\}$ means the second generation after the nth generation (e.g., grandchild), and so on. This single-variable difference equation represents the simplest nonlinear dynamical system.

To prevent any misunderstanding, three points must be made. First, the quantity $\{x\}$ is the normalized population and thus varies within the range $0 < x < 1$. Second, the parameter may or may not be normalized, depending on the proclivities of the user. Here it is not normalized, so that it can vary in the range $0 < a < 4$. Whenever $a > 1$, the population will increase, and whenever $a < 1$, the population will decrease. Third, although we talk in terms of populations and generations, we can use other variables such as dollars if we are dealing with budgetary changes, changes in the number of products manufactured, etc. In other words, we use the population as a surrogate for a variety of problems. The reason for this is because the logistic equation started out in the field of population dynamics and ecology.

Now for convenience let us multiply out the terms, and we get

$$x_{n+1} = ax_n - ax_n^2 \tag{2}$$

This is a simple quadratic equation wherein the first term on the right-hand side (r.h.s.) is linear, and the second term is nonlinear. Suppose that x_0, which denotes the initial normalized population, is a very small number such as 0.01. Further, let us assume arbitrarily that the parameter $a = 2$. Then the second term on the r.h.s., namely ax^2, = 0.0002. This is very much smaller than the first term on the r.h.s., namely, ax_n = 0.02, so that x_{n+1} and ax_n are approximately equal. Hence under such circumstances the nonlinear term can be neglected. This is basically what we mean by linearization.

For purposes of comparison let us next assume that the initial condition $x_0 = 0.5$, but that the parameter is untouched so that $a = 2$. If we introduce these values into Eq. 2, and perform the calculations, we find that $x_{n+1} = (2)(0.5) - (2)(0.5)^2 = 1 - 0.5$. Now the second term on the right side, i.e., the nonlinear term, is no longer negligible, and linearization is not justifiable. This is an example of being "sensitive to initial conditions," a cardinal characteristic of chaos. We shall revisit Eqs. 1 and 2 when we deal with the details of the discrete logistic equation. In the meantime, please remember that the values of x_{n+1} depend on both the parameter $\{a\}$ and initial value $\{x_0\}$.

The advantage of linearization is primarily mathematical in nature. Thus if we break up the equation representing a complex system into a number of linear equations without compromising the character of the system, the linear equations can be added to yield a solution that describes the total system. In contrast, adding nonlinear equations does not yield a solution to the problem. In many real-life problems linearization is not permissible because it is the higher-order or nonlinear terms that drive the problem. Hence, deleting them hastily could be like throwing out the baby with the bath water. Therefore, we must be very mindful about the initial values of $\{x_n\}$ and the value of the parameter $\{a\}$. The latter has a lot to say about the behavior of the equations. In applying chaos theory one does not normally linearize the equations. We shall return to Eqs. 1 and 2 when we discuss the discrete logistic equation at greater length, because it serves as an excellent model for instability and chaos.

In the pre-chaos theory another approach to solving nonlinear equations was to identify a suitable transformation whereby the nonlinear equation is transformed into a linear one. It is also possible to confine the solution of the problem to a region where it is asymptotic to a linear algorithm. With the advent of the computer, a very powerful tool became available and a new branch of mathematics was born. However, there is one aspect where computer solutions lack robustness. Specifically, computer approaches have trouble with singularities.

Chaos theory provides us with new techniques that grant insights into nonlinear problems without demanding that revolutionary breakthroughs in nonlinear mathematics per se must be discovered first. Of course, that does not mean that we should deny the importance of nonlinear mathematics. On the contrary: I recommend that chaologists build on it. Excellent starting points would be the vintage essay by von Kármán,[9] and the volumes by Beltrami,[10] and by Jordan and Smith.[11] It is not my intent to make contributions to nonlinear mathematics. Rather I wish to explore how we might gain insights into the behavior of nonlinear dynamic phenomena. We shall consider several techniques that have been gaining considerable attention. I shall cite them here only very briefly, leaving their detailed discussion to subsequent chapters. Because these are independent approaches I mention them in no particular priority or chronological order. One approach is due to A. M. Lyapunov (1857–1918), wherein one places the emphasis on understanding the stability of the problem. Admittedly this is more qualitative in nature than is an analytical solution, but it is very useful. Another approach that is proving to be very useful is the nonlinear integro-differential equation due to A. J. Lotka (1880–1949) and V. Volterra (1860–1940). For systems that are suspected to be chaotic, the relatively recent discovery of the so-called strange attractors offers exciting new vistas into nonlinear dynamics. Still another recent development that is proving helpful in the study of complexity is fractal geometry. All of these have found wide acceptance because of the ready availability of computers, particularly desktop computers.

The previous techniques are applicable because it is not necessary to describe nonlinear phenomena with nonlinear equations. Some of the most important nonlinear problems of physics are explained on the basis of linear equations.

Nonlinearity per se is neither desirable nor undesirable. Consider the age-old problem of turbulence. In process equipment it can lead to improved heat transfer, but it can also cause disastrous aircraft accidents when it is not controlled properly. In nature we encounter turbulence in the atmosphere and in the seas. It is one of the major unsolved problems of physics.

A typical example of turbulence is a flag crackling in a stiff breeze, rather than fluttering serenely in the laminar flow of a gentle breeze. Traditionally the transition from laminar to turbulent flow is determined by the critical value of the celebrated Reynolds number, which is a dimensionless parameter made up of the flow velocity $\{V\}$, the fluid density $\{\rho\}$, the fluid viscosity $\{\mu\}$, and the characteristic length $\{L\}$. The Reynolds number $\{Re\}$ is defined as follows:

$$Re = \frac{\rho VL}{\mu} \tag{3}$$

In this equation the numerator represents the inertia forces, while the denominator represents the viscous forces. This dimensionless ratio physically represents the ratio of the inertia to the viscous forces. Thus if the denominator is large, the flow will be sticky and slow, whereas if the denominator is small, the flow will be more rapid. Experimental evidence has shown that the higher the Reynolds number, the more likely it is for the flow to be turbulent because the viscous effects, i.e., the stickiness of the fluid, do not predominate. Generally, when $Re > 2,000$, the flow will be turbulent, and it will be laminar when $Re < 2,000$.

The transition from laminar to turbulent can become complex if there is an object in the flow such as an airplane wing, a propeller, or a turbine blade. This is shown in Fig. 5 from the famous Feynman lectures in physics.[12] The sketches show fluid flow around a cylinder, increasing Reynolds numbers proceeding downwards.

Depending on the magnitude of the Reynolds number, the flow can assume a number of very different configurations. At low Reynolds numbers, we note in Fig. 5b the formation of regular vortices, which in Fig. 5c are seen to start breaking off as the Reynolds number is increased. This is the celebrated von Kármán vortex street. Ultimately, in Fig. 5e, the flow assumes a turbulent boundary layer.

Not all of the transitions occur in a well-mannered, ordered way, and some materialize suddenly. This is a manifestation of the nonlinear nature of the problem. The drag forces acting on the cylinder increase or decrease depending on the Reynolds number. This is why an airplane can unexpectedly find itself in dangerous conditions.

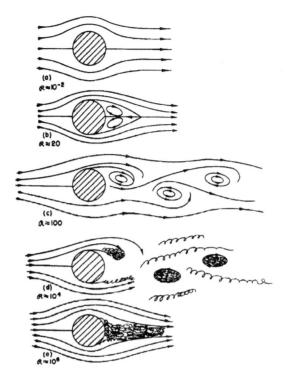

Figure 5. Flow past a cylinder at different Reynolds numbers. (R. P. Feynman, R. B. Leighton, and M. Sands (1964) *Feynman Lectures on Physics, Vol. III*. Copyright (©) 1964 by Addison-Wesley Publishing Company, Inc. Reprinted with permission of the publisher.)

A SNEAK PREVIEW OF CHAOS THEORY

Over the years I have found that when visiting a strange city, I can learn more about it if I first undertake a guided tour and then revisit by myself the particular sights that had intrigued me during the guided tour. Learning about a new subject is not too different. Accordingly, in the next few pages I shall expose the reader to a sneak preview of *chaos theory*.

Chaos theory is one way of studying complexity. Because there is such great variety among complex systems, it is reasonable to expect that there will be different methods of approach. One can pursue complexity along one of three paths: the spiritual, the philosophical, and the scientific. In this volume the emphasis will be on the scientific. This too can be pursued along different paths. I chose chaos theory because in my considered opinion it serves as the most powerful and general paradigm for the study of complex systems. The chaos theory approach that I shall follow here is an integrated interpretation of nonlinear dynamics, nonequilibrium thermodynamics, information theory, and fractal geometry. I shall discuss these as each makes its entrance in different parts of this volume. For the sake of fairness I mention in an endnote[13] some other approaches to complexity.

There is no one standard starting point to explain chaos theory. It is a heterogenous amalgam of different techniques of mathematics and science. Systems that upon analysis are found to be nonlinear, nonequilibrium, deterministic, dynamic, and that incorporate randomness so that they are sensitive to initial conditions, and have strange attractors are said to be *chaotic*. These are necessary but not sufficient conditions. For a system to be chaotic, the Lyapunov exponent must be positive. We shall get to this later.

It should be noted that there is no cause-and-effect relationship between the deterministic and random aspects. As we saw, the initial condition has an impact on the outcome of the events. Stated differently, the sensitivity to initial conditions causes the equations to be unstable. It is this instability that leads to uncertainty. The observation that deterministic equations may have unpredictable results is the essence of chaos. Deterministic chaos is not just arbitrary randomness: there is order involved. Thus, while chaos theory is unable to make long-term predictions, it can provide guidance for short-term predictions (Farmer and Sidorowich[14]) under certain circumstances when the attractor has only a few dimensions.

The sensitivity to initial conditions can be of utmost importance. It has been conjectured by Stephen Hawking,[15] the Lucasian Professor of Mathematics (the same chair that Isaac Newton once held) at Cambridge University, that had the density of the universe one second after the big bang been one part in a thousand billion greater, the universe would have collapsed after 10 years. However, had the density been smaller by the same amount, the universe would have been substantially empty after 10 years. Clearly a critical density was necessary for us to be around to read this book.

At the present it is believed that chaos can occur in three different types of systems: (a) conservative systems (MacKay and Meiss[16]); (b) dissipative systems; and (c) quantum systems (Gutzwiller[17]). In this volume I shall concentrate on chaos that is likely to occur in dissipative systems, because these are closest to real-life problems. Nevertheless, it is beneficial to realize that chaos is a specific, technically defined condition, quite different from its vernacular version. Also, please note that not all complex systems are chaotic.

Let us direct our attention briefly to the etymology of "chaos." It is generally held that scientific discoveries outrace social awareness. This is not always so. It is believed that the use of the word "chaos" goes back to the eighth century B.C., in Hesiod's *Theogeny*. In Part I of this work may be found:

> At the beginning there was chaos, nothing but void, formless matter, infinite space.

Later, in Milton's *Paradise Lost* we read:

> In the beginning, how the heav'ns and earth rose out of chaos.

We find "chaos" mentioned in Shakespeare's *Othello*:

> But I do love thee! and when I love thee not, Chaos is come again.

In the *Black Spring* Henry Miller wrote:

> A chaos whose order is beyond comprehension.

In some of the previous literary statements there is the implication that chaos is bad or undesirable. Usually, in our daily conversations we condemn chaos as some sort of confusion or disorganization. Scientifically, we look at it quite differently. Chaos implies the existence of unpredictable or random aspects in dynamic matters, but it is not necessarily bad or undesirable— sometimes quite the contrary. For example, the concept of self-organization pioneered by Ilya Prigogine, the 1977 Nobel Prize winner in chemistry, asserts that we can get "order out of chaos." The urbanologist Jane Jacobs[18] describes how during the industrial revolution, Manchester, England, deteriorated because it was orderly, indeed regimented, with its large ponderous factories, and did not have the flexibility to be competitive or the ability to adjust to a changing economic environment. In contrast, Birmingham, England, was quite disorganized with many diverse businesses. Birmingham was ". . . a muddle of oddments," according to Ms. Jacobs. Thus it was able to adjust to changing circumstances. It continues to thrive to this day, and the quality of life is high.

The American essayist and historian Henry Adams (1858–1918) expressed the scientific meaning of "chaos" succinctly: "Chaos often breeds life, when order breeds habit." Adams realized that ferment can be healthy and invigorating, while order can lead to stagnation. Nor is equilibrium always desirable. Lasers, which are so beneficial in communications, surgery, and technology, emit their coherent light when they are far from equilibrium. In this state things are in synchronism, and cooperation reigns.

It appears that the term "chaos" entered the scientific literature in 1975 when Li and Yorke[19] published a paper entitled "Period Three Implies Chaos," in which they characterized certain flows as being "chaotic." In an important 1976 paper, the eminent mathematician–biologist Robert May[20] pointed out that certain apparently simple equations may represent very complicated dynamics, and he referred to the Li–Yorke paper. Because of its importance, May's paper was widely read and it contributed to the acceptance of the term "chaos."

Sometimes "chaos" and "chaos theory" are confused with one another, and they are used interchangeably. This is like using the terms "quantum" and "quantum mechanics" interchangeably. This interchangeable usage can provoke misunderstandings. Chaos theory is a collection of mathematical, numerical, and geometrical techniques that allows us to deal with nonlinear problems to which there are no explicit general solutions. It should be noted that chaotic nonlinear equations are not general and have unique solutions for given initial conditions (Holmes[21]). Not all of the members of this coterie of techniques are new. What is new is a new paradigm that synergistically weaves together well-proven techniques with a more realistic outlook and with the benefit of the ubiquitous computer. Chaos theory is not a law in the manner that we consider classical mechanics, relativistic mechanics, thermodynamics, electrodynamics, or quantum physics. Not yet at least. The importance of chaos theory is its pragmatic *Anschau,* perhaps its *chutzpa,* to tackle problems that in the past we simply shied away from. Also, because of its generality, chaos theory can be used to analyze a variety of problems, even when these are nonchaotic. Maybe as the popular writer Robert Fulghum[22] refers to it, we should speak of "chaos science." Finally, it may have been with tongue in cheek that Zeldovich, Ruzmaikin, and Sokoloff[23] proposed that dynamic events that combine both determinism and randomness be called *divinamics,* after the ancient Roman *divinatio,* as described by Cicero.

As opposed to chaos theory, chaos is a condition. We noted earlier the specific conditions that must be obeyed for chaos to occur. Not all nonlinear dynamic problems are chaotic, but all chaotic problems are nonlinear. (At least we think so at this moment.) There is no one way that leads to chaos, and it can develop in numerous ways (Holmes[24]). However, within different classes of phenomena, chaos may be described best in accordance with what Eckmann[25] calls "scenarios." For example, in the case of turbulence— the premier manifestation of chaos—there are three major scenarios: the Ruelle–Takens–Newhouse scenario, the Feigenbaum scenario, and the Pomeau–Manneville scenario. Among these I shall rely greatly on the Feigenbaum approach of bifurcations.

Chaos and fractals have become buzzwords. They are mentioned in television shows, there are wall calendars of fractals, and there are coffee mugs. How this new science developed is narrated in the best-selling volume by the science writer James Gleick.[26] Chance and chaos are interpreted masterfully in a philosophical and scientific way by David Ruelle.[27] The outreach of chaos to literature is described by Katherine Hayles[28] of the University of Iowa in her scholarly volume. The immense, indeed the overpowering, popularity of the term "chaos" is unfortunate because it gives rise to confusion between its scientific meaning and its use in the vernacular. Nevertheless, all of this popularity is beneficial because it engenders a broad public awareness for a scientific field that does touch everybody. Hopefully, scientists and nonscientists will collaborate in solving common problems. The great Hungarian–American engineering scientist Theodore von Kármán noted that scientists try to discover what *is*, while engineers try to design what *isn't*. To this may be added that humanists try to discern truths, social scientists project how the institutionalization might take place, and artists give us inspiration. We all need to pool our efforts if civilization is to progress. As we proceed we shall note that efforts are being made to relate chaos to all of the aforementioned diverse fields. Perhaps chaos theory and its applications could serve as the common bond among different disciplines and serve as the means for mutual understanding, which is so badly needed.

DEGREES OF FREEDOM AND NUMBERS

Frequently, a system that has more components than another is considered to be more complex. Perhaps it is more sophisticated, perhaps the more the number of components, the more the probability that something will go awry, or perhaps it takes a longer time to construct. A system composed of many components can be expected to have a higher number of degrees of freedom. A monatomic molecule has fewer degrees of freedom than a polyatomic molecule. This is shown in Fig. 6 (Çambel[29]). At normal temperature and pressure, a diatomic molecule such as oxygen has five degrees of freedom, three translational ones in the x, y, and z directions, and two rotational degrees of freedoms. As the temperature is increased, additional degrees of freedom set in, such as the vibrational mode. As the temperature is raised further, the molecular bond can break

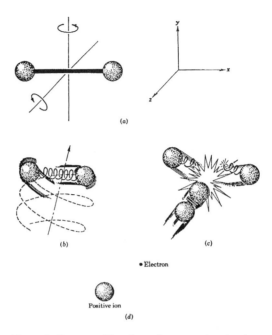

Figure 6. Degrees of freedom of atoms and molecules.

so that the oxygen molecule is dissociated into two oxygen atoms. Each of these has only three translational degrees of freedom. It is evident that the behavior of system is not defined completely by its own level of complexity, but also by the conditions it is exposed to, namely, its environment.

A paradox in our observations consists in the very large and very small numbers that we encounter. The number of cells making up the human body is estimated to be of the order of one hundred trillion, i.e., 10^{14}. Clearly, this is a fantastically large number. But numbers alone cannot tell the entire story. For example, there are about 2.7×10^{19} molecules in a volume of 1 cm^3. It follows that the number of molecules in even an average-sized room will constitute a horrendously large figure. Yet this large number of molecules of air cannot do any of the amazing functions that the cells in our bodies perform.

The intricacy of numbers that are involved becomes even more impressive if we note that a single cell contains billions of DNA steps, which are compacted into a microscopic space. A gene is made up of thousands of DNA steps, and in turn genes by the thousands make up chromosomes, which determine characteristics such as skin, eye or hair color, as well as physical handicaps. It is evident that the original cell has evolved into a far more complex structure, demanding a far higher level of organization; in other words, it can be self-organizing.

Scanning electron microscopy (McMahon and Bonner[30]) shows a single fertilized frog's egg splitting into 2 identical cells, then becoming 4-celled, 8-celled, and 16-celled. Then cell differentiation occurs so that some cells evolve into brain cells, others into skin cells, still others into blood cells and other parts and organs. Finally, parts and organs combine to form some biological system such as the frog or the human being. Once differentiation occurs, the different parts and organs assume different structures and perform different functions. It has been pointed out by R. May[31] that while we do not accurately know the number of living species, they must be counted in the millions, with different species exhibiting different levels of complexity.

Size

Some systems are extremely large, while others are minutely small. For example, our characteristic dimension,[32] e.g., our height, is of the order 1.75 m, but we live on Earth, which has a characteristic dimension of 12,756,000 m, its diameter at the equator. For convenience we can write this as $O\,(10)^6$ where the symbol O means the order of magnitude. In other words, the diameter of the earth is of the order of 10^6 m. In contrast, the characteristic dimension of DNA is exceedingly small, of the order of 10^{-8} m. The radius of the electron's orbit is even smaller, namely, 10^{-10} m. One of the amazing aspects of complex systems is how large and small systems can coexist.

According to Princeton University biology professor J. T. Bonner,[33] who has studied different life forms, complexity increases with size in both plants and animals. Extensive data going back to 3.5 billion years indicate that there is a tendency for organisms to increase in size, although the increase is not monotonic for the same species.

DYNAMICAL SYSTEMS

As a rule, complexity occurs in dynamical systems, namely, systems whose internal microscopic or external macroscopic motion is affected by one or more forces. Dynamical systems[34] may be *conservative* (Hamiltonian), i.e., they experience no energy losses. Conversely, systems can be *dissipative,* which is the case in most real-life situations that involve losses. Dissipative systems must receive energy, and information, in order to survive. Dynamical systems may also be classified as being *deterministic* or *stochastic.* We shall focus on the former.

Complexity is not limited to organic systems, but occurs widely in inorganic natural systems, in man-made systems, and in socioeconomic institutions. Particularly, the latter can grow and prosper, or conversely they can succumb to competing market forces.

Not all complex systems are self-organizing, but all self-organizing systems are complex. While we do not completely understand the evolution of biological systems, we realize that there is a difference between organic and inorganic systems. Jacques Monod,[35] the 1965 Nobel laureate in physiology and medicine, calls the characteristic of having a purpose, such as the preservation and multiplication of the species, "teleonomic."

The observation that complex phenomena can occur in relatively simple systems has far-reaching implications in many fields of physical science, chemistry, biology, and in different aspects of the social sciences, such as sociology and economics, and in the humanities, including history and philosophy. The arts, too, are involved. It may seem puzzling that so many widely differing fields have commonalities, encouraging intellectually promiscuous dialogue. Here the flow of information cannot be unidirectional, for neither the artist, the humanist, nor the scientist has any monopoly on knowledge. The discussion must have the benefit of contributions from all conceivable fields of learning. Purists who might object to such pragmatic, multidimensional thought transfer will hopefully take to heart the admonition of anthropologist R.N. Adams:[36] "There is nothing wrong in taking concepts developed in one science to see how they work in another if they yield an apt description." We must learn to learn from one another, and we must be heuristic.

SCOPE

Because complexity is so pervasive, it is necessary to have a theoretical basis for its understanding. There is much to the adage attributed to the 19th-century physicist and physiologist Ludwig von Helmholtz that there is nothing more practical than a good theory. As yet there is no fundamental law for complexity the way we conceive of the fundamental laws of classical mechanics, thermodynamics, electrodynamics, relativistic mechanics, and quantum physics. However, a number of very powerful approaches are being developed to enable us to deal with complex systems. Not all of these can be compressed into one small volume.

I wish to comment on why establishing scientific laws for complexity demands a nontraditional attitude. Basically, it is the challenge of reconciling the coexistence of determinism and chance in the same open structure. In chaos theory, determinism and chance are like the two sides of the same coin, and there is no cause-and-effect relationship between the two.

During the eighteenth and nineteenth centuries in particular, classical mechanics was tremendously successful, as it remains to this day. However, we now know that, like all scientific laws, Newton's laws apply only within certain simplifying assumptions. An example of such simplifications is neglecting the gravitational influence among planets that leads to the tidal effects. The fact that Newtonian mechanics endures is because it does describe many of our daily activities, whether this be mounting stairs, riding an elevator, lifting an object, or opening a door, etc. The fact that the computer I am writing this manuscript on processes information in accordance with modern physics does not change the fact that the computer itself remains at rest on my desk in accordance with classical physics.

It was generally believed that classical Newtonian mechanics is inadequate only in the cases of both very small particles where quantum mechanics is appropriate and very large bodies where relativistic mechanics applies. The recent developments in the study of complexity introduce new considerations.

It is possible to categorize physics in different ways. One possible classification is: (a) classical physics; (b) modern physics, i.e., relativistic and quantum physics; and (c) the sciences of complexity. An alternate approach that has been suggested by Davies[37] is: (a) the small; (b) the large; and (c) the complex. I prefer the first classification because complex systems can involve the small and the large concurrently, such as occurs in our visible bodies, which are made up of microscopic cells as well as macroscopic organs and in which take place some of the most intricate electro–chemical–biological processes.

It is ironic that our newfound ability to deal with complex systems was actually fertilized before modern physics was established. It started with the 1890 paper[38] by French mathematician Henri Poincaré, wherein he showed that Newton's laws did not provide a general solution in the case of the interplanetary system of the earth, the moon, and the sun. This is generally known as the *three-body problem*. What is so impressive is that a system that contains only three components, and hence would appear to be quite simple, is actually quite complex. What Poincaré was so prescient about was his realization that systems with few degrees of freedom believed to have been deterministic may upon further scrutiny exhibit irregular dynamics. In essence Poincaré was the progenitor of what we now call chaos. He described it as follows:

> ... it may happen that small differences in the initial conditions produce very great ones in the final phenomena. A small error in the former will produce an enormous error in the latter. Prediction becomes impossible, ...[39]

Why did it take some eight decades for Poincaré's work to be appreciated? Some reasons come to mind: First, imbued with the success of Newtonian mechanics, people were reluctant to abjure the sense of security that Laplacian determinism provided. This attitude still prevails. Second, at that time, because of the power of mathematical analysis, the popularity of geometrical studies was descending. I believe that there was still another reason. In his research Poincaré was concentrating on conservative, or Hamiltonian, systems. On the other hand, most dynamical systems, whether animate or inanimate, are dissipative and require energy inputs for subsistence. Last but not least, there were no computers in Poincaré's time.

It is the objective of this book to study complex systems in an interdisciplinary manner, but with the principal aid of chaos theory. Of course, in the final analysis, there is no one theory for all seasons, and frequently it is advisable to apply several approaches in a complementary manner. This leads to intellectual ethnicity, and hence to misunderstandings among different terminologies. The matter is further exacerbated because publications appear in widely scattered sources. None of this should discourage the chaoticist. As Ilya Prigogine allegedly remarked, "Nature is too rich to be described in one language."

QUO VADIS? REDUCTIONISM AND HOLISM

The traditional approach to scientific research has been reductionist. In other words, one successively breaks the system into its smaller and smaller components in an attempt to understand the fundamental structure and the inner workings. It is doubtful that the reductionist approach by itself will be fruitful in the understanding of complex systems. Let us listen to a person no less than the Nobelist Murray Gell-Mann:[40]

> I've spent most of my career working on the most basic level, that of the fundamental laws of physics . . . . To what extent is the so-called reduction of

each level of scientific description to a more basic level possible? When it is possible, to what extent is it a good strategy to pursue?

Clearly, there can be no straightforward, simplistic answer to this profound query. Physical chemist and social philosopher Michael Polanyi[41] suggests two levels of reality: the upper level represented by most any kind of machine, and the lower level by the parts of the machines. In the words of Polanyi:

> . . . the upper of these two levels is in fact unspecifiable in terms of the lower. Take a watch to pieces and examine, however carefully, its separate parts in turn, and you will never come across the principles by which a watch keeps time.

In some ways a complex system may be likened to a large mosaic tableau made up of a lot of little ceramic pieces. To the viewer the individual pieces mean little, but the entire mosaic can be a breathtaking image if the individual pieces have been properly shaped and correctly assembled.

Much as an antireductionist, holistic approach is tempting, there is the danger of pretending to be above the minutia and thereby becoming superficial. Perhaps the type of holistic approach that we should seek is in the spirit of the great British mathematician–philosopher Alfred North Whitehead,[42] who argued that in order to be excellent in one field, one must be at least as good in all of the related fields. In that spirit I intentionally did not name this section reductionism *versus* holism, nor reductionism *or* holism. If we are to understand complex systems better than we do now, it is imperative that we be willing to grapple with the details at the risk of being called reductionist, and that at the same breadth be able to rise above the tyranny of what is too restrictive and assume an holistic stance. We must change our hidebound ways of clinging to the past when it no longer serves reality. We must accept the challenge posed to us by the Austrian lyric poet Reiner Maria Rilke (1875–1926), who asked:

> Who's turned us around like this, so that we always, do what we may, retain the attitude of someone who's departing?

NOTES AND REFERENCES

1. Simon, H. (1957). *Models of Man*, New York: John Wiley & Sons.
2. Pagels, H. (1988). *The Dreams of Reason*, New York: Simon and Schuster.
3. Garfunkel, S., and Steen, L. A. (1988). *For All Practical Purposes—Introduction to Contemporary Mathematics*, New York: W. H. Freeman and Company.

4. Kolata, G. (1990). "In Shuffling Cards 7 Is Winning Number," *The New York Times/Science Times*, January 9, 1990.
5. Ford, J. (1983). "How Random Is a Coin Toss?" *Physics Today*, April, pp. 40–47.
6. Kline, M. (1953). *Mathematics in Western Culture*, New York: Oxford Univ. Press.
7. According to M. Kline (*op. cit.*), p. 210, Napoleon Bonaparte chided Laplace for not mentioning God in connection with his work on the functioning of the universe. Laplace evidently replied that he had no need for such a hypothesis.
8. Thompson, J. M. T., and Stewart, H. B. (1986). *Nonlinear Dynamics and Chaos*, Chichester, England: John Wiley & Sons.
9. von Kármán, T. (1940). "The Engineer Grapples with Nonlinear Problem," *Bull. Amer. Math. Soc.* **46**, 615–683.
10. Beltrami, E. (1987). *Mathematics for Dynamic Modeling*, Boston: Academic Press.
11. Jordan, D. W., and Smith, P. (1988). *Nonlinear Ordinary Differential Equations*, Oxford: Oxford Univ. Press.
12. Feynman, R. P., Leighton, R. B., and Sands, M. (1963). *The Feynman Lectures on Physics*, Reading, MA: Addison-Wesley Publishing Co.
13. In 1972, the French mathematician René Thom suggested a geometric or topological approach for the study of complex systems undergoing discontinuous changes. This approach is called "catastrophe theory" because according to Thom a system can be described by one of seven elementary catastrophes to which he ascribed romantic names like the "swallowtail catastrophe," the "butterfly catastrophe," the "cusp catastrophe" and others. At least one advantage of catastrophe theory is that by its geometric nature it provides a means for conceptualizing the dynamics of complex issues where no clear-cut data are available. For example, attempts have been made to apply it to decision making, management problems, and canine aggression. Those inclined to learn about the applications of catastrophe theory would benefit from the volume by Poston and Stewart.

There is nothing that is inherently undesirable about complexity. On the contrary, more often than not, complexity is the very cause of marvelous things because the various parts of the system "cooperate" to produce entirely different structures performing different functions. This was pointed out by Professor Hermann Haken the leader of the Stuttgart School, where the science of synergetics was founded. Haken defines the field as follows:

> Synergetics, an interdisciplinary field of research, is concerned with the cooperation of individual parts of a system that produces macroscopic spatial, temporal or functional structures. It deals with deterministic as well as stochastic processes.

This definition points out the breadth and depth that the study of complexity requires.

A revolutionary concept—that of cellular automata—was conceived in 1950 by John von Neumann based on ideas of Stanislaw Ulam. Looked at simplistically, a cellular automaton is a set of cells arranged in computer space. These automata change according to predetermined rules. One particular advantage of cellular automata is that they constitute discrete dynamical systems amenable to treatment with computers, whereas complex systems modeled with nonlinear differential equations are difficult to deal with. The current prominence of cellular automata is due to the contributions of Edward Fredkin, T. Toffoli, and Stephen Wolfram.

By now we are aware that complex systems entail incomplete knowledge and imprecision. One approach to this is through "fuzzy logic," developed by Professor Lotfi Zadeh of the University of California (Berkeley) in 1965. Fuzzy logic is a generalization of classical logic. Although at first a controversial subject, fuzzy logic is now being applied to management decision making, and it is being incorporated into the design of sophisticated equipment such as automatic cameras and home appliances.

A recent trend in studying complex systems is what Per Bak and his associates call "self-organized criticality."

14. Farmer, J. D., and Sidorowich, J. J. (1987). "Predicting Chaotic Time Series," *Physical Review Letters*, **59** (8), 845–848.
15. Hawking, S. (1991). "The Future of the Universe," *Engineering & Science*, **LV**(1), 13–21.
16. MacKay, R. S., and Meiss, J. D., compilers, (1987). *Hamiltonian Dynamical Systems*, Bristol, England: Adam Hilger.
17. Gutzwiller, M. C. (1992). "Quantum Chaos," *Scientific American*, **266**(1), 78–85.
18. Jacobs, J. (1969). *The Economy of Cities*, New York: Random House.
19. Li, T.-Y., and Yorke, J. A. (1975). "Period Three Implies Chaos," *American Mathematical Monthly*, **82**, 985–992.
20. May, R. M. (1976). "Simple Mathematical Models with Very Complicated Dynamics," *Nature*, **261**, 459–467.
21. I am grateful to Professor Philip Holmes of Cornell University for pointing this out to me in a personal communication dated March 3, 1992.
22. Fulghum, R. (1988). *It Was on Fire When I Lay Down on It*, New York: Ivy Books. *Note:* I am grateful to Lois Meesen for bringing this reference to my attention.
23. Zeldovich, Y. A. B., Ruzmaikin, A. A., and Sokoloff, D. D. (1990). *The Almighty Chance*, Singapore: World Scientific.
24. Holmes, P. (1984). "Bifurcation Sequences in Horseshoe Maps: Infinitely Many Routes to Chaos," *Physics Letters*, **104A**, 299–302.
25. Eckmann, J.-P. (1981). "Roads to Turbulence in Dissipative Dynamical Systems," *Reviews of Modern Physics*, **53** (4), 643–654.
26. Gleick, J. (1987). *Chaos*, New York: Viking.
27. Ruelle, D. (1991). *Chance and Chaos*, Princeton, NJ: Princeton University Press. *Note:* A review of this volume prepared by this author may be found in *Nonlinear Science Today*, **2**, 1992.
28. Hayles, N. K. (1990). *Chaos Bound—Orderly Disorder in Contemporary Literature and Science*, Ithaca, NY: Cornell Univ. Press.
29. Çambel, A. B. (1963). *Plasma Physics and Magnetofluidmechanics*, New York: McGraw-Hill Book Co.
30. McMahon, T. A., and Bonner, J. T. (1983). *On Size and Life*, New York: Scientific American Books, Inc.
31. May, R. M. (1988). "How Many Species Are there on Earth?" *Science*, **241**, 1441–1455.
32. Objects have different dimensions such as height, width, and depth. But they have other discernible characteristics such as mass, temperature, or energy level. Any of these can be considered as a useful metric to describe the system under consideration. The generic term that is used is *characteristic dimension*. There is no standard characteristic dimension. Instead of using the height of a person in the example in the text, we could have chosen the shoe size, but that would have been less indicative in comparing the size of different persons. Common sense is a good guide.
33. Bonner, J.T. (1988). *The Evolution of Complexity*, Princeton, NJ: Princeton Univ. Press.
34. While motion is inherent in both dynamic and kinematic systems, in kinematic systems no forces are involved.

35. Monod, J. (1971). *Chance and Necessity*, New York: Alfred A. Knopf, Inc. Reissued as a paperback in 1972 by Vintage Books, New York.
36. Adams, R.N. (1988). *The Eighth Day*, Austin: Univ. of Texas Press.
37. Davies, P., (ed.) (1989). *The New Physics*, Cambridge, England: Cambridge Univ. Press.
38. Poincaré, J. H. (1890). "Sur le Problème des Trois Corps et Les Equations de la Dynamique," *Acta Mathematica*, **13**, 1–270.
39. This quotation is abstracted from "Chaos," by J. P. Crutchfield, J. D. Farmer, N. H. Packard, and R. S. Shaw (1986). *Scientific American*, **255**, p.48.
40. Gell-Mann, M. (1988). "Simplicity and Complexity in the Description of Nature," *Engineering and Science*, **LI**, 2–9.
41. Polanyi, M. (1959). *The Study of Man*, Chicago: Univ. of Chicago Press.
42. Whitehead A. N. (1929). *The Aims of Education and Other Essays*, New York: The MacMillan Company.

CHAPTER **2**

META-QUANTIFICATION OF COMPLEXITY

INTRODUCTION

So far we have viewed complexity in a conceptual manner. Hence, the discussion has been substantially qualitative. Now it is appropriate to ask, "How complex is complex?" The question is not just a matter of academic curiosity, but is of considerable practical importance. We must all make decisions: the policy maker involved in political–socioeconomic issues, the physician making choices among alternate methods of treatment, the researcher working on the frontiers of science, the taxicab driver in choosing the best route to get you to the airport during heavy traffic time, and everybody else. Each of these situations involves uncertainty. In turn, uncertainty is related to complexity. Therefore it would be dandy to have a scale whereby one could measure complexity. Unfortunately, the complexity of a system is not an intrinsic property of the system, but is a composite

27

of a variety of factors, some of which were discussed in the previous chapter. In addition, complexity is what Casti[1] calls a "contingent" property of the system which is a very apt statement.[2]

Measures for composite factors have been concocted. While some are useful, others can be misleading. For example, a few years ago I was listening to the weather report on TV when I heard the term "humiture." I called the station to inquire what the term meant and was informed that it was the sum of the temperature and the humidity. I explained that this was meaningless because most any pair of temperature and humidity could add up to the same number yet have entirely different physiological properties. The station stopped using the term "humiture." The moral is that ascribing a scale say from 0 to 10 to measure complexity is meaningless. On the other hand, we cannot just shrug our shoulders and give up. Indeed there are a number of partial measurements. That is why I called this chapter "Meta-Quantification." Later, in Chapters 7, 8, 9, and 10, we shall take up various indicators of chaos. While not all complex systems are chaotic, many are, so that the metrics of chaos such as fractal dimension, Lyapunov exponential coefficients, Kolmogorov–Sinai entropy, and others can help us gauge the degree of complexity. I shall follow with caution the adage of Galileo Galilei: "Measure what is measurable, and make measurable what is not measurable." Let us start with some cautionary statements by a number of thoughtful persons.

FACING THE NEW REALITIES

It is quite evident that the arrogance of certainty espoused by classical physics is no longer valid, and the Laplacian paradigm no longer holds. Three major discoveries contributed to this demise. The work of Albert Einstein shattered the belief in the existence of absolute space time. The uncertainty principle of Werner Heisenberg established that there can be no accurate measurements. Last but not least, Kurt Gödel's undecidability principle (we shall consider this later in this chapter) showed that there can be no absolute and final proof. Thus there can be no absolute free will because of interconnectedness and uncertainty. Nevertheless we still labor under the influence of Lord Kelvin,[3] who insisted:

> When you can measure what you are speaking about, and express it in numbers, you know something about it, when you cannot express it in numbers, your knowledge is of a meager and unsatisfactory kind; it may be the beginning of knowledge, but you have scarcely, in your thoughts, advanced to the stage of science.

Practitioners in the hard and social sciences are accustomed to justifying their conclusions by measured or calculated results. This is justifiable in traditional situations that are well documented through previous observations and where standard measurement techniques are available. The situation is different in the case of complex systems, where there are all sorts of imponderables. As a rule, complex systems are not in equilibrium, whereas most measurements are made in the equilibrium state. Also, because complex systems are open they communicate with other systems, and this additional information will prejudice the results of measurements. More important, we do not have a good understanding of complex systems. After all, only about two decades have passed since we started to study them systematically. We cannot explain them with confidence, and we even have trouble describing them. To make useful measurements one must know what one is measuring. As yet we do not have this luxury when dealing with most complex systems. That thoughtful medical researcher and great humanist Lewis Thomas[4] commented as follows:

> The task of converting observations into numbers is the hardest of all, the last task rather than the first thing to be done, and it can be done only when you have learned, beforehand, a great deal about the observations themselves. You can, to be sure, achieve a very deep understanding of nature by quantitative measurement, but you must know what you are talking about before you can begin applying the numbers for making predictions.

We are not sufficiently advanced to express the levels of complexity in terms of numbers as Lord Kelvin would like us to do, and perhaps we never shall be.

In the absence of something concrete, it is not unusual to rely on qualitative or geometrical descriptions in lieu of numerical data. Richard Feynman, the eminent theoretical physicist and Nobel laureate, foresaw this in 1963 when chaos theory was not a household term. He ended the chapter on fluid dynamics in his celebrated lectures[5] with the prescient words:

> The next era of awakening of human intellect may well produce a method of understanding the *qualitative* content of equations. Today we cannot. Today we cannot see that the water flow equations contain such things as the barber pole structure of turbulence that one sees between rotating cylinders. Today we can not see whether Schrödinger's equation contains frogs, musical composers, or morality—or whether it does not.

I have shared the above quotations because I want to emphasize that we have to set aside our penchant for numbers. Numbers are in the realm of facts, not truths, and there is a great difference between the two. It is not unlike the difference between precision and accuracy. An instrument may be very precise, yet give inaccurate measurements because it is incorrectly located within the system.

Admittedly, a variety of unconventional techniques is being developed for quantifying complexity, but the jury is still out on their merits. The task is formidable, due to the nature of complexity. Here I shall survey some of the paths that have been followed. These are the hierarchical, geometric, and analytical approaches.

HIERARCHICAL APPROACH

Looking at complexity as a hierarchical sequence implies the evolution of complexity. For example, Bonner[6] has shown that in animals and plants the larger ones tend to be more complex. Prigogine[7,8,9] and his associates have looked at it from the viewpoint of self-organization, whereby transitions can occur from chaos to order. The quantification of complex behavior is discussed in the volume edited by Mayer-Kress.[10] Different measures of complexity have been brought together in the volume by Peliti and Vulpiani.[11] In this section we discuss measures of complexity based on knowledge and information content.

The concept of a hierarchy of learning was proposed by the French philosopher–mathematician Auguste Comte (1798–1857), who argued that more complex bodies of knowledge depend on relatively simpler ones. One would start with mathematics, proceed through astronomy to physics, and then to chemistry, followed by biology. Finally, there would be "sociologie" (sociology), a term coined by Comte,[12] and one that today would most likely embrace all of the social sciences and their applications.

The hierarchical levels of Comte should not be interpreted as meaning that if you know physics, then you know chemistry. Rather, what he meant was that to formulate an understanding of chemistry one must know physics, but physics by itself is not sufficient to explain chemical reactions; it is necessary to build on physics the structure that constitutes the corpus of chemistry. Physicist B. West and virologist J. Salk[13] (the discoverer of the polio vaccine) proposed two hierarchies for complexity, shown in Fig. 1. One is complexity in nature, and the other is complexity in knowledge. Further, they proposed three stages: physical sciences, life sciences, and human sciences, in ascending order. The human sciences would include the decision-making process and would involve a great amount of complexity

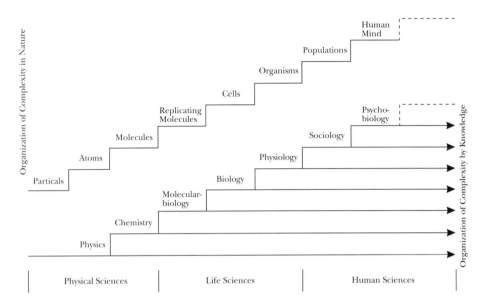

Figure 1. The evolution of complexity in nature and knowledge according to B. West and J. Salk. (Reproduced with the permission of the authors.)

and uncertainty. The term "human sciences" includes all disciplines and interdisciplines that in one way or another are related to quality of life of the human species. Because science and technology play such an important role in our lives, it is obvious that the physical and life sciences lead up to the human sciences. It appears that we have come full circle to the earlier viewpoint that was known as natural philosophy and science.

Complexity is inherent in social and political structures such as families, institutions, cities, the nation-state, and alliances. Regardless of whether these entities are large or small, in general, decision making and governance are becoming more complex as a result of the mounting number of driving forces, such as increasing populations, new technologies, and rising human expectations. It has been pointed out by philospher–engineer Frederick Koomanoff [14] that this has resulted in an increase in the types of formal and informal organizations that share governance. Koomanoff's interpretation of how the scope of governance has had to change by increasing populations and technological innovations is shown in Fig. 2.

As may be seen, 100 years ago the state of technology brought forth producers and users; 50 years later governments became involved; about 20

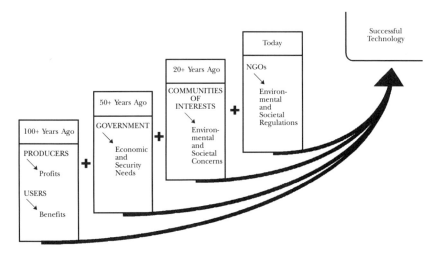

Figure 2. Evolution of interorganizational complexity according to F. A. Koomanoff. (Reproduced with the permission of the author.)

years ago it became necessary to be mindful of community interests, while today nongovernmental organizations have become influential in policy formulation and governance. Of course, each level of organization or grouping has various subsets, which have also been proliferating.

Technological advancement, too, follows a hierarchical restructuring. This process appears to be guided primarily by the intelligence of techno-logical innovation. The information content, $\{I\}$, and the material content, $\{M\}$, of various manufactured goods were compared by Fritsch,[15] Çambel,[16] and by Çambel and Fritsch,[17] who suggest that the complexity in the manufac-turing sector is related to the I/M ratio of the products. As demands increase, and resources become scarce, there is a tendency for the $\{I/M\}$ ratio to rise, which is shown in Fig. 3.

In this industrial phase space one starts with handcrafts in the lower left-hand quadrant, Q-1, where the I/M ratio is quite low. For a while the manually produced items grow in size, such as the heavy industries (Quadrant-II), but eventually in Quadrant-III an increase in information is required, such as in modern hospitals or high technology weapon systems, both very complex. Computers with their ever more powerful microchips fall into Quadrant-IV. Of course, there is overlapping of the different levels of industry among developed and developing nations. Further, it is not necessary for any one region to remain in the same quadrant. As the

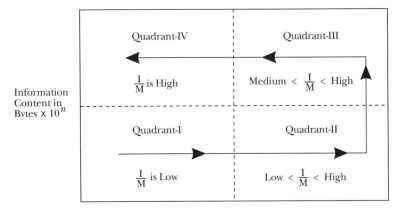

Figure 3. Industrial phase space according to A. B. Çambel and B. Fritsch.

history of industrialization indicates, some nations go through the historical steps, some leapfrog, while others lag.

It might be conjectured what lies beyond Q-IV. I doubt that this is the final stage. Historically, each succeeding revolution has helped the preceding one survive. The industrial revolution has bolstered the agricultural revolution, and the computer revolution is making it possible for the industrial revolution to become more productive. If Q-IV is to survive, it can be expected that there will be a more advanced stage. A case can be made that this will be one where human values are cherished.

GEOMETRIC APPROACH

Complexity can manifest itself both in form and in function, and frequently in both. The Japanese art of origami, which requires that one start with a simple, plain, square sheet of paper and, through a series of folding steps, arrive at an intricate figure is an example of evolving complexity. Figure 4 from Kasahara[18] shows the steps leading to a paper jet plane that is entirely different from the original sheet of paper. Both its shape and function have changed, i.e., it can be used as a maneuverable toy. The original sheet of paper after undergoing the origami transformation has become more complex. Whereas originally it was a two-dimensional plain sheet of paper,

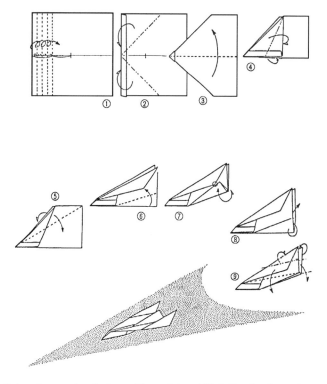

Figure 4. Origami paper jet airplane. (From *Origami Made Easy* by K. Kasahara. Copyright (©) 1973. Reprinted with permission of Japan Publications, Inc.).

its structure and dimensions have become intricate. No longer is the origami figure a two-dimensional surface. It has acquired depth, namely some additional dimensionality. Robert Lang[19] has aptly titled a paper of his "Origami: Complexity Increasing."

We shall see in a later chapter when we discuss the geometry of fractal shapes that the origami jet plane has a noninteger dimension. We shall return to the geometric evaluation of complexity through the application of fractal dimensions after we learn about strange attractors associated with chaos, which consitute one of the paradigms for complexity. Such fractal shapes are best constructed with the assistance of computers. Various degrees of geometric complexity in natural and artificial objects and images generated by computer graphics may be seen in two imaginative books by the biophysicist and computer visualizer par excellence Clifford Pickover.[20,21]

ALGORITHMIC COMPLEXITY

Last but not least, we mention the mathematical way of measuring complexity known as *algorithmic complexity*. An algorithm is a precisely defined, step-by-step computational procedure that is not ambiguous. The game plan is to model a complex system that has associated with it a large number of data points by establishing the minimum length of the algorithm that describes the arrangement of the numbers of the data set. The mathematical methodology constitutes the *SKC theory*, which stands for the initials of Solomonoff,[22] Kolmogorov,[23] and Chaitin,[24] who worked on it independently in the mid-1960s. Their algorithmic complexity theorem provides a means of measuring randomness that, in turn, gives us information about the extent of the complexity. The SKC theory may be applied to characterize complex systems that are between complete randomness and complete determinism. It follows that the model of a well-determined sequence such as {AAAAA} would be short and simple because a single instruction would suffice, in this case "type {A} five times." In contrast, if the letters of the alphabet are arranged randomly, e.g., {AYUGL}, many instructions would be necessary. Because of the expanding theoretical and practical importance of this concept in a variety of fields, I shall digress to provide a brief background. I should note that the subjects of randomness, computability, and algorithmic complexity are by no means fully understood. Fortunately, Gregory Chaitin continues the research in the field and is its acknowledged guru.

In their monumental work, *Principia Mathematica* (1910–1913), philosopher–mathematicians Bertrand Russell and Alfred North Whitehead propounded the argument that all mathematics can be deduced from a set of mathematical axioms. German mathematician David Hilbert [25] (1862–1943), after whom Hilbert space is named, and English mathematician F. P. Ramsey [26] (1904–1930), known for having shown that complete disorder is impossible, were of the view that there can indeed be procedures to establish whether a proposition follows from a set of axioms. Their weighing in on the side of the Russell–Whitehead argument was a major endorsement. However, in a seminal paper in 1931, mathematician–logician Kurt Gödel [27] (1906–1978) proved that within a formal system questions exist that are neither provable nor disprovable on the basis of the axioms of that system. In doing so he showed that there are problems that cannot be solved by any set of rules or procedures because this would always require a higher set of rules. This is known as "Gödel's undecidability theorem or incompleteness theorem." (It should not be confused with Heisenberg's uncertainty principle.[28]) There was considerable consternation when Gödel first expounded his theorem. Not only did it upset the status quo, but it was also difficult to comprehend. However, things have settled down. Furthermore, a number of straightforward proofs of the theorem have been developed. A masterful review has been provided by Chaitin,[29] to whom we owe so much for his contributions to algorithmic information theory.

In his prize-winning book,[30] Douglas Hofstadter described how Gödel's mathematical discovery is an extension of the musical compositions of J. S. Bach (1685–1750) and the art of the Dutch artist M. C. Escher (1902–1972). The applications of Gödel's mathematical theorem are far-reaching. It reoriented the thinking in classical logic, primarily with the work of Alonzo Church. Its greatest impact was on computers and artificial intelligence. For example, Chaitin[31,32] has discussed how it relates to information and mathematical reasoning. These matters are of great importance in complex systems and also are related to entropy, which is a measure of randomness and chaos, which, of course, bring us back face to face with complexity. We shall consider the different implications of the entropy function in a later chapter.

The undecidability principle gained particular importance in a number of applications, such as artificial intelligence. Alan Turing[33] (1912–1954) showed that certain problems cannot be decided in a finite number of steps on universal Turing machines.[34] This abstract machine can simulate any computer. It follows that a random number generator does not generate truly random numbers. Not only did Turing conceive a universal computer, and extend Gödel's work, but in doing so he made it possible to establish algorithmic theory, which at the moment is probably our most powerful way of evaluating complexity.

We noted that an algorithm should not leave any room for uncertainty. A number is said to be computable if a computer algorithm can arrive at it easily. For example, consider the string of numbers {11111}. The computer algorithm would be quite simple, namely, the statement to print {1} five times. Similarly, the algorithm for the string of numbers {1234567891011-12131415} would be the statement to print the numbers from 1 through 15 sequentially. Even the number 0.538461539 is computable because it is the decimal form of the fraction 7/13 and hence has a simple algorithm.

There is no unique algorithm for computable numbers. For example, the number 100 can be represented by a number of algorithms, such as: 10^2, or $110 - 10$, or 2×50, or $1,000/10$. Random numbers, on the other hand, are noncomputable in the sense that the algorithm may be as long as the number, or is the number itself. The algorithm for the numbers obtained by rolling an unweighted die would be random and hence as long as the number of times we threw the die. In this context, Chaitin and Kolmogorov looked for the length of the minimal program. They argued that the complexity of a string of digits is the length of the shortest program that a Turing machine would have to have in order to produce that string. The numbers constituting the minimal program must be random. The shorter the string or the algorithmic program, the less complex the problem.

If we know the mechanism whereby the string of numbers was derived, we can tell whether or not we are dealing with randomness. For example, regardless of the appearance of a string, knowing that it resulted from the honest tossing of an unweighted coin means that there was randomness. If a coin is tossed 10 times, there will be any of 2^{10} binary sequences. Let the string of heads and tails be {HTHTHT}. This does not look random, but it is because it resulted from a random event. Let us next be given a string of numbers {101010101010101010} without being informed how it was derived. It does not look random, but we cannot be sure.

Next let us consider a string that came from a cache of red (R) and white (W) stones and does look random, for example, {RRWWWRWRRWRRWW-RRRWWWRRR}. If the stones came from a well-mixed urn, we would know that the string is random. However, a string appearing random may not be random. As Rasband[35] has pointed out, we might have been told to place the red stones below a certain nonlinear curve, specifically a logistic map, and the white stones above it. Because the generating process is deterministic the string is not random.

Important applications of algorithmic complexity are in the coding of DNA molecules, information theory, and artificial intelligence. It is a powerful metric for characterizing complexity.

None of the previous techniques is sufficient to set the level of complexity of a system. Other techniques that pertain to chaos are useful and should be used. We shall consider these in later chapters.

NOTES AND REFERENCES

1. Casti, J. (1984). "System Complexity," *Options—IIASA*, **3**, 6–9.
2. The dictionary definition of "contingent" is "1. dependent for existence, occurrence, character, etc., on something not yet certain; conditional. 2. liable to happen or not; uncertain. 3. happening by chance or without known cause." See Flexner, S. B., and Hauck, L. C. (1987). *The Random House Dictionary of the English Usage*, second edition, New York: Random House.

3. Thomson, W. (Lord Kelvin) (1891–1894). Popular Lectures and Addresses. London.
4. Thomas, L. (1983). *Late Night Thoughts on Listening to Mahler's Ninth Symphony*, New York: Viking Press.
5. Feynman, R. C., Leighton, R. B., and Sands, M. (1964). *The Feynman Lectures on Physics*, Volume II (41–2). Reading, MA: Addison-Wesley Publishing Co.
6. Bonner, J. T. (1988). *The Evolution of Complexity by Means of Natural Selection*, Princeton, NJ: Princeton Univ. Press.
7. Prigogine, I. (1980). *From Being to Becoming*, New York: W. H. Freeman and Co.
8. Prigogine, I. and Stengers, I. (1984). *Order out of Chaos*, Toronto: Bantam Books.
9. Nicolis, G., and Prigogine, I. (1989). *Exploring Complexity*, New York: W. H. Freeman and Co.
10. Mayer-Kress, G. (1986). *Dimensions and Entropies in Chaotic Systems*, Berlin: Springer-Verlag.
11. Peliti, L., and Vulpiani, A., eds. (1988). *Measures of Complexity*, Berlin: Springer-Verlag.
12. Edwards, P., ed. (1967). *The Encyclopedia of Philosophy*, **2**, 173–177, New York: Macmillan Publishing Co., Inc.
13. West, B. J., and Salk, J. (1987). "Complexity, Organizations, and Uncertainty," *European Journal of Operational Research*, **30**, 117–128.
14. Koomanoff, F. (1992). "Energy, the Enabler, in a Changing World," *Nuclear Engineering and Design*, (In Press.)
15. Fritsch, B. (1981). *Wir werden überleben*. Munich: Günter Olzog Verlag.
16. Çambel, A. B. (1984). *An Exploratory Study of the Impacts of Technological Innovation*, Report to EPRI, #RP-2416-07-01.
17. Çambel, A. B., and Fritsch, B. (1987). "A Synergistic Approach to Evaluate Technological Change," *Alternative Energy Sources VII*, T. N. Veziroglu, ed. **6**, 17–37. Washington: Hemisphere Publishing Co.
18. Kasahara, K. (1973). *Origami Made Easy*, Briarcliff Manor, NY: Japan Publications, Inc.
19. Lang, R. J. (1989). "Origami: Complexity Increasing." *Engineering & Science*, **XII** (2), 16–24.
20. Pickover, C. A. (1990). *Computers, Pattern, Chaos and Beauty*, New York: St. Martin's Press.
21. Pickover, C. A. (1991). *Computers and the Imagination*, New York: St. Martin's Press.
22. Solomonoff, R. J. (1964). "A Formal Theory of Inductive Control," *Information Control*, **7**, 224.
23. Kolmogorov, A. N. (1965). *Doklady Akademii Nauk*, **119**, 754; and "Three Approaches to the Quantitative Definition of Information," *Problems of Information Transmission*, **1**, 1.
24. Chaitin, G. J. (1966). "On the Length of Programs for Computing Binary Sequences," *Journal of the Association of Computing Machinery*, **13**, 547; also see Chaitin, G. J. (1975). "Randomness and Mathematical Proof," *Scientific American*, **232**, (5), 47–52.
25. Newman, J. R. (1956). *The World of Mathematics*, New York: Simon and Schuster.
26. Graham, R. L., and Spencer, J. H. (1990). "Ramsey Theory," *Scientific American*, **263**, (1), 112–117.
27. Gödel, K. (1931). "On Formally Undecidable Propositions of Principia Mathematica and Related Systems," (title translated), *Monatshefte für Mathematik und Physik*, **38**, 173–198.
28. The uncertainty principle in quantum mechanics enunciated by Werner Heisenberg in 1927 is conceptually quite different from Gödel's undecidability theorem. The uncertainty principle sets limits on the accuracy with which the positions and momenta of particles can be specified. If either is measured accurately, there will be uncertainty in the measurement of the other. The degree of this uncertainty is fixed by Planck's constant.
29. Chaitin, G. J. (1982). "Gödel's Theorem and Information," *International Journal of Theoretical Physics*, **21**(12), 941–954.

30. Hofstadter, D. R. (1979). *Gödel, Escher, Bach: An Eternal Golden Braid,* New York: Basic Books.
31. Chaitin, G. J. (1982). *op. cit.*
32. Chaitin, G. J. (1988). "Randomness in Arithmetic," *Scientific American,* **259**(1), 80–85.
33. Turing, A. (1937). "On Computable Numbers with an Application to the Entscheidungs Problem," *Proc. London Mathematical Soc.,* **42**, 230–265.
34. Since Turing's original work in 1937 *(op. cit.),* a number of advances and modifications have been made, and there is a rich literature on the subject. We mention here a few that the inquiring reader might locate easily: Dewdney, A. K. (1989). *The Turing Omnibus,* Rockville, MD: Computer Science Press; Penrose, R. (1989). *The Emperor's New Mind,* Oxford: Oxford University Press; Ralston, A., ed. (1983). *Encyclopedia of Computer Science and Engineering,* New York: Van Nostrand Reinhold Company; Rucker, R. (1987). *Mind Tools,* Boston: Houghton Mifflin Co.; Zurek, W. H. (1990). *Complexity, Entropy and the Physics of Information,* Reading, MA: Addison-Wesley Publishing Co.
35. Rasband, S. N. (1990). *Chaotic Dynamics of Nonlinear Systems,* New York: John Wiley & Sons.

CHAPTER **3**

THE ANATOMY OF SYSTEMS AND STRUCTURES

INTRODUCTION

One need not be particularly perceptive to note that complexity is not uniform and that it appears in a wide variety of forms. The study of complexity is the "melting pot" of widely differing disciplines; accordingly, semantics lose parochialism, and new idiolects appear. Sometimes the same term means different things, and sometimes different terms are used to mean the same thing. Perhaps it is best to heed Austrian philosopher Ludwig Wittgenstein: "Don't look for the meaning, look for the use."

OPEN, CLOSED, AND ISOLATED SYSTEMS

It is helpful to speak of "systems" without having to elaborate on all of the details. By definition, a *system* is a collection of organic or inorganic matter, or even of institutional entities, surrounded by a wall. This wall may be real or imaginary, just as you may choose to erect a fence around your property or leave it unbounded. Even if there is no actual fence, people have some clues as to where your property boundaries are. In the same spirit we talk of different types of walls surrounding complex systems. Depending on the type of wall, three different types of systems are defined: isolated systems, closed systems, and open systems.

Isolated systems are surrounded by walls so rigid and impermeable that nothing can cross them. In real life there are no true isolated systems, because there is always some sort of leakage. Isolated systems cannot survive for long. As the lyric poet John Donne (1573–1631) wrote:

> No man is an island, entire of itself, every man is a piece of the continent, a part of the main; . . . .

The concept of isolated systems is used in scientific work because by excluding external forces such as gravity and electromagnetism the analysis can be more tractable and provide useful insights, albeit at best only approximate. In economics, too, externalities are excluded, and this also is unrealistic because economic systems are influenced by government actions, social demands, and political forces. In reality, systems will be pulled or pushed away from equilibrium and they will behave nonlinearly.

Closed systems are surrounded by walls that do not allow the passage of matter, although they allow energy and information to pass through them. In the context of our discussion *open systems* deserve elaboration. These have permeable walls so that matter, energy, and information or entropy may cross them in either direction. The human body is a good example of an open system. We eat nourishing food, we drink liquids, and we eliminate waste. We receive energy and we expend it. We obtain information, we process it, and we pass it on. In the past, attempts were made to explain living systems by means of the thermodynamics of isolated systems. That these attempts were unsuccessful should be evident at once.

In the realm of nonliving open systems consider the automobile. We fill the fuel tank of our automobiles, and the exhaust products come out from the tailpipe—both exchanges of matter. Energy is exchanged because the fuel contains chemical energy that is expended as mechanical energy at the wheels. Traffic lights provide information to the driver, and taillights inform other drivers about one's intentions.

Institutions too are open systems. All sorts of goods are brought in and out of buildings. People come in and go out and thus contribute to both material and information exchanges. Of course, telephone lines or satellite dishes help exchange information between the institutional building and affiliates outside.

These are common examples of organic, inorganic, and institutional open systems. They are also dynamic, and they differ in relative size. Technically, the flows (fluxes) may be matter, energy, and information, or its variation—the entropy. In the vernacular, the flows would be called exports or imports of goods or services. The characteristics of open and isolated systems are compared in Table I.

Table 1. Characteristics of systems.

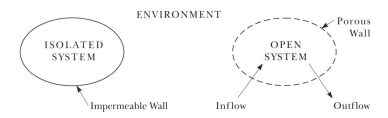

1.) Adiabatic or perfectly insulating wall separates the system from its surroundings. Absolutely nothing can cross the wall in either direction.	1.) A permeable wall separates the system from its surroundings. Hence there are flows (i.e., in and/or out) across the wall. The flows include material, energy, information, entropy.
2.) We only consider the macroscopic aspects of the system.	2.) We consider both macroscopic or microscopic aspects of the system.
3.) The system is in equilibrium.	3.) The system may or may not be in equilibrium, and this can occur on either the macroscopic or microscopic levels. It may be near to equilibrium (linear case), or it may be far from equilibrium (nonlinear case).
4.) The system undergoes only reversible processes, i.e., it is conservative, Hamiltonian.	4.) The system can undergo irreversible processes, i.e., it is dissipative.

PHASE SPACE

We are accustomed to dealing with two-dimensional planes, e.g., the floor we walk on, or three-dimensional space as we move from floor to floor. It is customary to specify the position of a point in rectangular Cartesian coordinates having x, y, and z directions. This is insufficient when dealing with dynamic systems because they do not stay put in one place, but move around, i.e., the position is time-dependent. We express this by writing $[x(t)$, $y(t)$, $z(t)]$. Lest there is any doubt about this assertion, imagine watching a flock of birds, or a school of fish, and attempting to identify them individually. They constantly change place and vary their velocity. Hence both the position coordinates and the velocity (or momentum) coordinates must be specified. The dynamics, or motion, is then described by the following set of equations:

$$\frac{dx}{dt} = \dot{x} = f_1 (x,y,z) \tag{1}$$

$$\frac{dy}{dt} = \dot{y} = f_2 (x,y,z) \tag{2}$$

$$\frac{dz}{dt} = \dot{z} = f_3 (x,y,z) \tag{3}$$

Here $\{f_1\}$, $\{f_2\}$, and $\{f_3\}$ are functions of the position coordinates $\{x\}$, $\{y\}$, and $\{z\}$, while $\{dx/dt\}$, $\{dy/dt\}$, and $\{dz/dt\}$ are the corresponding velocities. In mathematical shorthand these may be expressed by the letters $\{x\}$, $\{y\}$, and $\{z\}$ with dots above them.

This set of equations represents a third-order dynamical system. If x, y, and z are distances, their derivatives will be velocities. However, the state variables may be other entities such as population or economic indicators. Spaces made up of state variables and their derivatives are called *phase spaces*, whereas if state variables constitute the coordinates, one may be speaking of *state spaces*. The two terms are frequently used interchangeably, and this does not do any harm as long as we know with what we are dealing. The system is said to be *autonomous* if the functions $\{f_1, f_2,$ and $f_3\}$ do not depend on time explicitly.

Because the concept of phase space is grounded in statistical physics, let us take a minute to differentiate between its meaning in classical statistical physics and in modern dynamical systems. It turns out (Davidson[1]) that describing the events in a gas during a one-second time span would require

3×10^{35} entries. Even with large computers this would be an awesome task. Accordingly, one resorts to statistical techniques. One such abstraction is phase space, promulgated by the great American scientist Josiah Willard Gibbs (1839–1903). Typically in statistical physics the state of a system is specified by two generalized coordinates, the positions denoted by the generalized vector $\{q_i\}$, and the momenta denoted by the generalized vector $\{p_i\}$. Therefore, if the number of particles is denoted by $\{n\}$, the state of the system at any instant is given by $\{2n\}$ numbers. In three-dimensional Cartesian coordinates, one speaks of "6n-dimensional phase space." This approach is peculiar to Hamiltonian systems that have an even number of dimensions and apply to conservative systems. In contrast, real-life systems are dissipative, not conservative, because they experience losses.

The concept of phase space is used when dealing with complex systems, but it differs from the formulation in statistical mechanics (Jackson[2]). Systems consist of a variety of discrete elements. Their number $\{n\}$ does not need to be even, and one uses the concept of n-dimensional phase space. This is not a regular, symmetrical, and neat coordinate system as we are accustomed to, and could easily look as sketched in Fig. 1.

Although it is difficult to do so, this figure should be visualized as being n-dimensional, i.e., there is a ray for each dimension, $\{x_1, x_2, x_3, x_4, x_5 \ldots x_n\}$, however many dimensions it might take. Here $\{n\}$ is the number of real-valued variables that are involved in the problem. This type of phase space is more general than the one associated with conservative systems. It

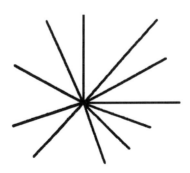

Figure 1. n-dimensional phase space.

follows that spaces can be constructed in different ways, depending on the demands of the situation.

We assume that there are suitable laws that govern the previous functions and that these laws are represented by the following ordinary differential equations:

$$\frac{dx_1}{dt} = f_1(x_1, x_2) \tag{4}$$

$$\frac{dx_2}{dt} = f_2(x_1, x_2) \tag{5}$$

The solutions to these equations form a set of "trajectories" or "flow lines" on the (x_1, x_2) phase plane that together with the trajectories is called the *phase diagram* (Seydel[3]). Alternatively, a group of trajectories may be called the *phase portrait,* and an arbitrary one is sketched in Fig. 2.

When a trajectory closes on itself so that it forms a loop, it is said to be an *orbit.*

As early as 1895 the founding father of dynamical systems theory, Henri Poincaré,[4] inquired:

> Does the moving point describe a closed curve? Does it always remain in the interior of a certain portion of the plane? In other words, and speaking in the language of astronomy, we have inquired whether the orbit of this point is stable or unstable.

This query brings us face to face with two concepts crucial to dynamical systems: equilibrium and stability.

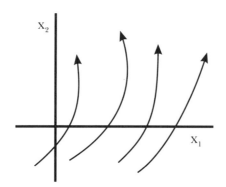

Figure 2. Phase portrait.

EQUILIBRIUM AND NONEQUILIBRIUM

Traditionally one considers the equilibrium state like a harbor of tranquility. Actually, life is a journey, not a destination. In the world of complex dynamical systems, it is frequently the equilibrium state that is like a temporary weigh station. Dynamic processes can take place only when the system deviates from equilibrium. Accordingly, it is helpful to comprehend the limitations of equilibrium.

The dictionary[5] defines "equilibrium" as "the state of rest or of balance due to the equal action of opposing forces." This does not mean that nothing is happening. For example, a chemical reaction may be in equilibrium because the reactants and products have equal, albeit opposing, rates and there is a balance. We commonly write this as follows:

$$\text{Reactants} \rightleftharpoons \text{Products} \tag{6}$$

Similarly, the thermostat that controls the furnace at home ensures thermal equilibrium by turning the furnace on and off to provide sufficient heating to make up for the heat lost to the outside.

Mathematically, the equilibrium state can be described quite tersely. Thus if $dx_1/dt = dx_2/dt = 0$, the system is in equilibrium. When the system comes to rest, one speaks of "fixed points." We shall examine these in greater detail in connection with "attractors." In certain instances "critical conditions" occur here, for example, bifurcations.

There are several different kinds of equilibrium. *Microequilibrium* occurs when the smallest constituent (microsystem) of the system is in equilibrium with its immediate surroundings. Because the observable behavior of macroscopic systems is due to the microscopic contributions, the term *macroequilibrium* applies to systems when their composition is considered in bulk, as is the custom in classical, equilibrium thermodynamics. In turn, *global equilibrium* occurs when a number of systems are in equilibrium mutually. Under certain circumstances, although the overall system may undergo an irreversible process, small parts may have internal parameters that are substantially constant over time. This is called *local equilibrium.* These different equilibria are related sort of like the bumper sticker that says, "Think globally, act locally."

It is a truism to assert that when there is no equilibrium we have *nonequilibrium.* The point is that there are different classes of nonequilibrium. In particular, we differentiate between two regimes of nonequilibrium. One is *near to equilibrium.* As a rule this applies to transport phenomena such as electric conduction, heat conduction, and mass transfer. These are commonly described by Ohm's law, Fourier's law, and Fick's law, respectively. In such near-to-equilibrium regimes, flows are linear functions of the forces (potentials). In contrast, in many complex systems the flows are complicated, nonlinear functions of the forces, and we speak of *far-from-equilibrium* conditions. We shall be primarily interested in these.

STABILITY AND INSTABILITY

Because complexity involves dynamics, it stands to reason that it occurs under nonequilibrium conditions. Equilibrium and stability are twin subjects. Just as we considered a variety of types of equilibrium, there are many different types of stability. It is beyond the scope of this volume to discuss all of these, and accordingly, I refer the reader to the erudite discussion by V. Szebehely.[6]

Instability can occur in all kinds of structures: solid, fluid, animate, inanimate, and institutional. Perturbations can render stable systems to be unstable. However, this is not all that bad, because from the unstable structure may arise more complex and better organized systems both in form and in function. Clearly then, stability is an important aspect of complexity. Figure 3 displays different configurations that are stable and unstable.

A system may be normally stable, but rendered unstable because of some perturbation. The type and magnitude of the perturbation as well as the susceptibility of the system are factors that must be considered. Frequently, more than one type of perturbation is necessary to cause the transformation from the stable to the unstable state. For example, it is believed that certain diseases can develop only when several factors conducive to them occur together. The probability that all of these will coexist cannot be very high, and chance must be crucial.

Another example is social unrest. In spite of numerous types of public dissatisfaction and frustration, major upheavals do not start unless a variety of ripe conditions is present. It was Lenin himself who said, "Not every revolutionary situation leads to a revolution." There was more than the lack

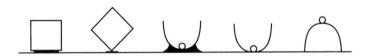

Figure 3. Simple cases of stability and instability.

of bread that sparked the French revolution! The failure of machines and crops are two further examples. A system may become unstable suddenly, but it is usually long in the making. It is like the proverbial last straw that broke the camel's back, a manifestation of nonlinearity. Furthermore, we see effects of chance and necessity daily around us, and we notice that different entities are affected differently by similar circumstances. Thus it is not uncommon for a stable system to move into an unstable state, which in turn can lead to a new stable order.

It is instructive to differentiate between "stable" and "unstable" equilibrium. If every initial state in phase space is close to equilibrium and moves on to other states that also are close to equilibrium, we say that the system is stable. But what does "close to equilibrium" mean? After all, systems in the equilibrium state may be so in varying degrees. Accordingly, we differentiate between equilibria that are "weakly stable" and "strongly stable."

If for any initial displacement

$$x(0), \quad \dot{x}(0), \quad \text{we have} \quad x(t) \to 0, \quad \dot{x}(t) \to 0, \text{while } t \to \infty$$

we have *asymptotic stability*. We shall encounter this when we discuss the logistic curve.

We know from experience that a flow cannot occur unless there is an imbalance of driving forces; there must be a potential difference or lack of equilibrium. A term related to the equilibrium state is the so-called steady flow situation. For the case when mass is conserved, we write a Liouville-type master equation as follows:

$$\frac{\partial \rho}{\partial t} + \nabla \cdot (\rho V) = 0 \tag{7}$$

where $\{\rho\}$ is the density and $\{V\}$ is the barycentric velocity, which is a weighted average of appropriate reference velocities. The interpretation of this equation can vary depending on whether the Eulerian viewpoint— where the observer is fixed in space—or the Lagrangian approach—where the observer travels with the flow—applies.

The "substantial" or "material" or "total" derivative is defined

$$\frac{D}{Dt} \equiv \frac{\partial}{\partial t} + V \cdot \nabla \tag{8}$$

where the first term on the r.h.s. is the local rate of change with time, and the second term on the same side is the spatial rate of change depending on position. The motion can be steady or unsteady even if $\partial \rho / \partial t = 0$, i.e., even if the density does not vary with time. However, in some publications this is also called the "stationary state," or "steady equilibrium," to differentiate it from static equilibrium. In summary, the motion is steady if $V = V(x)$ and unsteady when $V = V(x, t)$.

PARAMETERS TO EVALUATE EQUILIBRIUM

In a number of applications it is desirable to know the rate at which internal equilibrium is approached in an active system. We face this personally whenever we work out in the gym. The *recovery time* is the time elapsed for the raised heartbeat to return to its normal level. An entirely different application occurs during atmospheric reentry of space vehicles when the high kinetic energy is transformed into thermal energy as the vehicle enters the dense atmosphere. This causes the gases in the stagnation region to become dissociated or ionized. To ensure the structural integrity of the vehicle and to prevent communications blackout, the design engineer must know whether or not the chemically excited state will return to equilibrium. In such technological nonequilibrium situations it is helpful to define the *relaxation time* as the time it takes for an excited system to reach its equilibrium state. We denote this time by $\{t_{rel}\}$. Further, we assume that it takes the time $\{t_{tr}\}$ to travel over the distance under consideration. A dimensionless ratio $\{\Pi\}$ (called the Damköhler number) is defined as follows:

$$\Pi = \frac{t_{tr}}{t_{rel}} \tag{9}$$

If this ratio > 1, then there is sufficient time for the system to equilibrate, whereas if the ratio < 1, then there will not be sufficient time for equilibration; the system is said to be "frozen."

This concept of dimensionless parameters can be adapted to applications outside the physical–chemical sciences. The following have been suggested by Çambel[7] for the analysis of socioeconomic development:

$$\text{Education parameter} = \frac{\text{Time to acquire education}}{\text{Time for education to become outdated}} \tag{10}$$

$$\text{Engineering parameter} = \frac{\text{Time to commercialize innovation}}{\text{Time for product to become obsolete}} \tag{11}$$

$$\text{Leadership parameter} = \frac{\text{Time in office}}{\text{Time to effect change}} \tag{12}$$

In terms of the education parameter, one cannot help but be intimidated by the phenomenal rate of increase in knowledge. Hence the numerator should rise, but, even so, a person who is normally considered to be educated can easily become intellectually illiterate in one field or another. This suggests that the denominator of the education parameter will decrease with time. Whereas the numerator should be increased to ensure proper preparation, this is not feasible, as the cost of education keeps rising. There are two routes to deal with the dilemma: Either the educational system will have to be overhauled or individuals will have to return to school to update themselves. Probably both the educational system will have to change and reeducation will have to occur. The proliferation of short courses and seminars indicates that people are willing to undergo intellectual rejuvenation. In any case a sound and broad education will have to form the educational base if we are to behave as human beings rather than bionic robots.

The engineering parameter is a measure of technological virility and entrepreneurship. Some years ago, Ralph Cordiner, the respected CEO of the General Electric Co., stated that half of the products his company manufactured did not exist a decade earlier. Or, compare the items offered in the 1902 Sears, Roebuck catalog[8] with that of the present and you will note product obsolescence qualitatively and quantitatively. Technological innovation is also very evident. Perhaps more interesting is the change in social morés.

The life span of products depends on their kind. In general, high-tech items have greater turnover, partly because their manufacturers enjoy positive feedback, thereby creating new products. They feed on these. For example, the industrial revolution has helped sustain the agricultural revolution by making it more productive, and the information revolution that we are now in is helping its predecessor, the industrial revolution, to survive by making it more productive. New technologies also have the good fortune of being able to fill new markets. Concerning the numerator of the engineering parameter, the time for commercialization depends greatly on how well those in the R&D laboratory work with the marketing force, thereby attracting consumers. Further, the issue depends greatly on institutional incentives and barriers nationally as well as internationally. There is another matter that must be considered, namely, measuring the time for commercialization. Where does one start? For example, how long did it take for civilian nuclear power to become commercialized? The first U.S. nuclear reactor for the generation of electric power was opened in 1958 in Shippingport, Pennsylvania. We know that this was an experimental plant, and it was not commercially owned. However, let us not debate these matters, and let us allow it to serve as a benchmark of a plant dedicated to electric power generation. Now how long did it take to get there? How far should we go back for the starting point? Do we start with the Fermi reactor in Stagg Field that went operational in 1942, or should we go back to the nuclear fission work of Otto Hahn and Lise Meitner in the 1930s, or should we go back to Rutherford's early experiments with atoms? There is no one correct answer, particularly because technological acceptance depends on the demands of social institutions. As we saw in the previous chapter, Koomanoff has shown how societal governance changes, in part driven by technology, and in part driving it.

There are various factors that enter the leadership parameter. The time in office depends on the type of organization, its mission, the stage of maturity, the stability of the economy, and plenty of politics. In general, the leader should remain in place to see the progress initiated become accepted. If the time in office is too long, mental and ethical calcification takes root. On the other hand, if the time in office is too short, important initiatives may never be completed because the successor in office may have other priorities.

I must note emphatically that this type of nonequilibrium analysis can be helpful only if values, objectives, and priorities are carefully identified and articulated.

RAYLEIGH–BÉNARD INSTABILITY

Solids have a specific shape, while liquids occupy the vessel they are in, and gases fill the entire vessel if it is covered. But there is a more interesting structure that liquids can manifest. This is the celebrated Rayleigh–Bénard convection model,[9] and it may be found in fluids being heated from below. It was shown experimentally in 1900 by H. Bénard[10] that when a suitable fluid between two close horizontal plates is heated from below, a cellular structure is observed, which when viewed from the top appears like well-ordered cells that look like a honeycomb.

The experiment consists of a fluid between two horizontal plates, usually a few millimeters apart. The temperature of the lower plate $\{T_L\}$ is higher than that of the upper plate $\{T_H\}$, i.e., $T_H < T_L$. Originally the heating is by conduction, but as the temperature gradient increases, i.e., the departure from equilibrium increases, the hot water rises, encounters the colder strata, cools and tumbles over. It is these convection rolls that appear like hexagonal cells when viewed from the top. The reader would do well to feast his or her eyes with the exquisite photographs by M. G. Velarde, M. Yuste, and J. Salan in the beautiful album edited by Professor Milton Van Dyke.[11] The experiment may be conducted in apparatuses of different shapes. In Fig. 4 from Velarde and Normand[12] are shown the convection rolls when viewed from the side. The rolls may be in either direction. As Pierre Bergé, Yves Pomeau, and Christian Vidal[13] have shown, in an experiment free of imperfections the rolls may rotate in either direction once they reach a critical Rayleigh number that we discuss next.

In 1916 Lord Rayleigh presented the theoretical interpretation of the associated stability phenomena and showed that it depends on the dimensionless Rayleigh number $\{Ra\}$ named after him. This parameter is defined $Ra = g\alpha\beta d^4/\kappa\nu$. Here $\{g\}$ is the coefficient of acceleration; $\{\alpha\}$, $\{\kappa\}$, $\{\nu\}$ are the coefficients of expansion, thermal conductivity, and kinematic viscosity, respectively; $\{\beta\}$ is the temperature gradient; and $\{d\}$ is the distance between the plates. The Rayleigh number is one of the major criteria for the behavior of natural or free convection. Physically it represents the ratio of the buoyant forces to the viscous or dissipative forces. When it reaches a critical value, convection columns appear. The magnitude of the Rayleigh

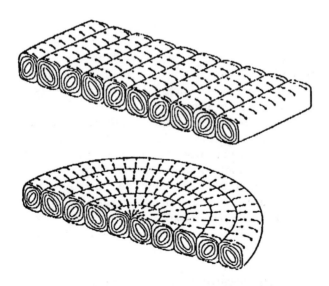

Figure 4. Rolls of convection cells. (From "Convection" by M. G. Velarde and C. Normand. Copyright (©) 1980 by Scientific American, Inc. All rights reserved.)

number determines the stability of the fluid behavior during natural convection. Below a critical Rayleigh number, Ra_{cr}, perturbations will die out, while above the Ra_{cr}, perturbations will tend to grow. Bergé, Pomeau and Vidal have pointed out that at the instability threshold of the Rayleigh number, a bifurcation occurs. This is why the rolls may rotate in either direction. Bifurcations are a symptom of instability, and we shall face this matter again when we get more deeply involved with chaos. The Rayleigh–Bénard instability has applications in meteorology and the process industries. Of course, the dimensions in such applications will be much greater than the mm–cm range generally used in laboratory settings. A good rule of thumb to remember is that the heated horizontal dimension should be much greater than the vertical distance between the two plates.

IRREVERSIBILITY

Whenever a system undergoes a process, some losses can be expected. Such processes are said to be *irreversible* because we cannot trace them backward without leaving some sort of evidence. Even a good craftsman will leave some marks when fixing a broken piece of china. Real-life situations are irreversible.

Consider the case of an automobile that is suddenly brought to rest by putting on the brakes. This is sketched in Fig. 5. The kinetic energy of the automobile is converted into thermal energy at the surface where the tires touch the pavement. During this process high-quality kinetic energy is degraded to low-quality thermal energy. This thermal energy spreads over parts of the pavement and the tire, where it is absorbed. The particles making up the tire and the roadbed become agitated as they now find themselves at a higher temperature, although imperceptibly so. The particles of the tires and the street pavement have been disordered, and we are faced with a dissipative effect. We ask if it might be possible to upgrade the energy, namely, collect all of the dispersed thermal energy and somehow convert it back to kinetic energy so that the car is set into motion without stepping on the gas pedal. This would indeed be a miraculous feat and is unlikely to happen because we are faced with an irreversible process which cannot retrace its path without any aftereffect whatsoever. The only way of getting the car moving again is to step on the gas pedal.

Let us consider another irreversible phenomenon. In this scenario the cap of a bottle containing a highly fragrant perfume is removed. From experience we know that after a while the perfume molecules will escape

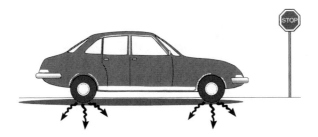

Figure 5. Automobile brought to a sudden stop.

from the bottle and the scent will permeate the air. Let us fictionally assume that the air particles and the perfume particles had been tagged with signs in a sufficiently visible manner so that we might record the events with a videotape recorder. Such a depiction would show the mixing of the perfume molecules with the air molecules. Let us run the tape first forwards and then backwards. The difference would be obvious at once, because we know that the perfume molecules are very unlikely to reenter the bottle. The genie has been let out of the bottle, and it will not go back in. We have a clear-cut irreversible process. Mixing processes are irreversible, as we all know well when one pours cream into a cup of coffee and stirs it.

The dispersion of the perfume in this example constitutes a mass transport phenomenon that is generally expressed by Fick's diffusion equation:

$$\frac{\partial c}{\partial t} = D \nabla^2 c$$

(13)

where c is the concentration, and D is the diffusion coefficient.

Let us perform another experiment. In this, we make a videotape of the periodic motion of the pendulum of a grandfather clock. The frictional and drag losses that would bring the pendulum to a stop are overcome by the clock mechanism. By experience we know that once set into motion the pendulum will oscillate between the same two extreme positions as long as the clock mechanism is wound up. If we run the videotape forwards and then backwards, the viewer would have no inkling in which direction it is being run unless he or she cheats and looks at the clock dial! The oscillatory motion of the pendulum appears to be reversible because the clock mechanism constantly makes up the energy lost in overcoming friction and drag. Systems that continue to function because they are nourished by energy addition are called "dissipative systems."

A mathematical shorthand way of differentiating between reversible and irreversible processes is by testing for time reversal invariance. Wave phenomena are expressed by the well-known wave equation:

$$\frac{\partial^2 V}{\partial t^2} = \nabla^2 V \tag{14}$$

where V is the velocity, t denotes time, and the so-called nabla operator, ∇^2,

$$\nabla^2 = \frac{\partial^2}{\partial x^2} + \frac{\partial^2}{\partial y^2} + \frac{\partial^2}{\partial z^2} \tag{15}$$

Let us inquire what happens to the wave equation and Fick's equation as we change the sign of the time $\{t\}$. Thus if one reverses the time in the wave equation, i.e., $t \Rightarrow -t$, the problem would obviously not be altered because time in the denominator is squared, i.e., $t^2 = (-t)^2$. However, changing the sign of the time in Fick's equation would definitely alter the problem because time is of first order. Reversible processes are time reversal invariant. The opposite, namely, time directionality, is a primary characteristic of complex systems.

The processes that real-life systems undergo are irreversible because they are accompanied by losses such as friction, mixing, heat transfer, and unrestrained expansion. For such systems to continue functioning they must receive sustenance in the form of energy. Structures that exist only as long as additional energy is supplied are said to be dissipative. Such structures may be found not only in applications of physics and chemistry, but also in biology and social organizations. Some dissipative systems are self-organizing because they can achieve higher levels of complexity. Life forms are self-organizing. In considering self-organization it is preferable to use the term *structure* because the original system changes over time and becomes more highly organized or structured. For example, the constituent particles of steam are in a state of highly agitated random motion, and any particle may be at most any one place at any one moment. By contrast, in a crystal the particles assume a prescribed arrangement; they have structure. Between the gaseous and the solid states are liquids. Their particles may have some semblance of arrangement, but only on a local basis, whereas in the case of crystals the arrangement is global, or they have coherence. Coherence is both space- and time-dependent. We shall return to this fascinating subject when we discuss the "E-word," namely, entropy.

NOTES AND REFERENCES

1. Davidson, N. (1962). *Statistical Mechanics*, New York: McGraw-Hill Book Co.
2. Jackson, E. A. (1989). *Perspectives on Nonlinear Dynamics*, Vol. 1., Cambridge, England: Cambridge Univ. Press.
3. Seydel, R. (1988). *From Equilibrium to Chaos*, New York: Elsevier.
4. Davis, H. T. (1962). *Introduction to Nonlinear Differential and Integral Equations*, New York: Dover Publications, Inc.
5. Flexner, S. B., and Hauck, L. C., eds. (1987). *The Random House Dictionary of the English Language*, New York: Random House, Inc.
6. Szebehely, V. (1984). "Review of Concepts of Stability," *Celestial Mechanics*, **34**, 49–64.
7. Çambel, A. B. (1984). "A Synergistic Approach to Energy-Oriented Models of Socio–Economic–Technological Problems," *Synergetics—From Microscopic to Macroscopic Order*, (pp. 183–196), E. Frehland, ed. Berlin: Springer-Verlag.
   Note: For stimulating discussions regarding these parameters, I am indebted to the late Dr. Milton Clauser of TRW, Inc., Professor Max Dresden of S.U.N.Y., Professor Bruno Fritsch of E.T.H., and Professor A. K. Oppenheim of the University of California (Berkeley).
8. Reissue of the *1902 Edition of the Sears, Roebuck Catalogue*, **MCMLXIX**, New York: Crown Publishers, Inc.
9. Rayleigh–Bénard convection differs from Marangoni convection in that in the latter the heating occurs from the top.
10. Bénard, H. (1900). "Les Tourbillons Cellulaires dans une Nappe Liquide," *Revue Générale des Sciences Pures et Appliquées*, **11**, 1261–71 and 1309–28.
11. Velarde, M. G., Yuste, M., and Salan, J. (1982). *An Album of Fluid Motion*, M. Van Dyke, ed. Stanford, CA: Parabolic Press.
12. Velarde, M. G., and Normand, C. (1980). "Convection," *Scientific American*, **243**, July, 92–108.
13. Bergé, P., Pomeau, Y., and Vidal, C. (1984). *Order within Chaos*, New York: John Wiley & Sons.

CHAPTER **4**

ATTRACTORS

INTRODUCTION

Phase space is the playing field of dynamic phenomena. Systems move in all sorts of directions, execute strange patterns, and sometimes stop. For example, a children's swing will come to rest in a vertical position unless it receives pushes. Wind up a yo-yo and let it unwind. It will wind itself back up for successively shorter distances and eventually come to rest at the lowest point allowed by the length of the string. Or consider a grandfather clock and its pendulum. The pendulum will come to rest at a stable vertical position when the clock mechanism stops. Such stable equilibrium points are called *fixed-point attractors*. The term "attractor" derives from the observation that if a system in phase space is near an attractor, it tends to evolve towards the state represented by that attractor. Dynamical systems are attracted to attractors the way fireflies are attracted to light. And just as a firefly can come from most any direction and be attracted to bright lights, so a system can start from different sets of points in phase space and still wind up at the same attractor region, called the *basin of attraction*. The totality of basins makes up the phase space.

The fixed-point attractor is one of four types of attractors that we observe in dynamical systems. The other three are *limit cycles*, *tori*, and *strange attractors*. Attractors play a very important role in nonlinear dynamics. Their configuration can tell us whether the system is conservative or dissipative; it can also help us figure out whether or not the system is chaotic. In some ways attractors may be looked upon as the geometric counterpart of the entropy function, which also tells us about the behavior of the system. We shall examine the various interpretations of the entropy function in a later chapter.

Contrary to a fixed point in phase space, limit cycles are represented by loops, which may or may not be closed. We shall discuss these in this chapter in connection with the van der Pol equation, and then in a later chapter in connection with the predator–prey model. Tori exhibit characteristics of limit cycles and strange attractors, but have their own peculiar behavior.

The attractor known as a strange attractor is an important indicator of chaos. It is worth noting that while fixed-point attractors, limit cycles, and tori are predictable, strange attractors exhibit unpredictable and bizarre motions. This is why they are also called *chaotic attractors*.

A good place to start in discussing attractors is the harmonic oscillator, which is well represented by a pendulum. Depending on the complexity there are different oscillators:

    a.   the undamped, unforced, linear oscillator
    b.   the undamped, unforced, nonlinear oscillator
    c.   the damped, unforced, linear oscillator
    d.   the damped, unforced, nonlinear oscillator
    e.   the damped, forced, linear oscillator
    f.   the damped, forced, nonlinear oscillator

While we shall consider these here, they are treated in detail by Thompson and Stewart.[1]

In addition to differentiating between real and ideal oscillators, we must ascertain whether they are linear or nonlinear. Of course, all pendulums swing along a curvilinear arc. What we really inquire about is whether the equation is linear or nonlinear in the mathematical sense. In general we deal with a linear situation if the angle of the arc is relatively small, and it is nonlinear when the angle of the arc is large. Consider the simple pendulum shown in Fig. 1 as our model.

In this figure we have a pendulum having a length $\{l\}$, with a bob having a mass $\{m\}$. We pull the pendulum away from its vertical equilibrium position

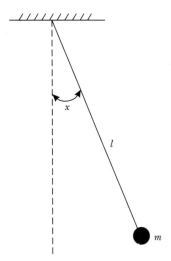

Figure 1. Pendulum.

by an angle $\{x\}$, and let it swing. The equation of motion is well known and generally written as follows:

$$ml\ddot{x} + mg \sin x = 0 \qquad (1)$$

where the $\{x\}$ with the double dots over it denotes the acceleration. Equation 1 is nonlinear and hence has no closed-form solution. However, we can linearize it for very small angles $\{x\}$, because for these $\{\sin x \cong x\}$. One can write as an approximation the following linear equation of motion:

$$\ddot{x} + Ax = 0 \qquad (2)$$

wherein $A = g/l$.

FIXED-POINT ATTRACTORS

The equilibrium state and the stability of the system are intimately related. Two pertinent questions follow: To what extent does a system return to its original equilibrium position after it has been disturbed, and further, is there any chance that the disturbance will cause the system to assume another orbit? If every state in the phase space initially close to equilibrium proceeds to other states all forever close to equilibrium, i.e., t → ∞, then the equilibrium is said to be stable.

The upper figure of Fig. 2 depicts a pendulum that oscillates between two extremes. The lower part of this figure shows the trajectory leading to a fixed point in the position–velocity plane. We first pull the pendulum to its farthest position from the vertical axis. While we hold it there its velocity will be zero, but its position will be at its maximum. Then we let the pendulum go and allow it to swing. As the pendulum swings back and forth, we plot both its position along the vertical axis and its velocity along the horizontal axis. From these we obtain the inward spiralling trajectory.

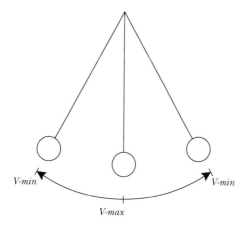

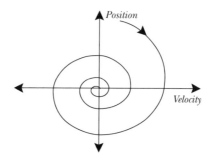

Figure 2.   Free-swinging pendulum (upper figure) and its position–velocity trajectory (lower figure).

With each swing, the distance of the pendulum from the vertical position decreases. This is due to the dissipative effects, namely the friction at the hinge of the pendulum, and the air drag acting on it. Eventually, the pendulum will come to rest at the vertical position or at its fixed-point attractor. No matter how far the pendulum was deflected at the beginning, it will always come to rest at this same fixed-point attractor. This means that a fixed-point attractor is not sensitive to initial conditions. Because sensitivity to initial conditions is one of the indicators of chaos, we conclude that fixed-point attractors do not lead to chaos.

In contrast to the trajectory of a pendulum experiencing losses, the trajectory of a lossless pendulum (an idealization) would be a closed, circular orbit in the position–velocity phase space, shown in Fig. 3.

The particular orbit would depend on the initial position, and hence two orbits are shown to emphasize the alternatives. This apparent dependence on initial conditions should not be construed as chaos, because the trajectories are closed. In the case of chaos, the end points of the attractors do not meet.

Returning to the lower part of Fig. 2 where the spiral is inward, this is only so when there is a stable fixed point. But things do get unraveled once in a while, and for such unstable situations the spiral will point outward. This can be noted in another type of structure in phase space, the *saddle point*, which is sketched in Fig. 4. Here the trajectories that are going towards the saddle point are stable, and those that are going away are unstable.

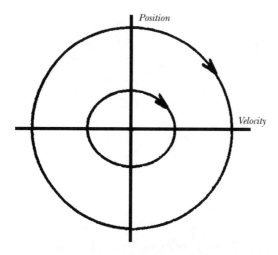

Figure 3. Orbits of lossless pendulum in phase space.

Figure 4. Saddle point.

LIMIT CYCLES

The second type of attractor is the limit cycle. We shall discuss limit cycles in two different ways: (a) the van der Pol oscillator, and (b) the predator–prey model. In this section I shall confine myself to the van der Pol oscillator, which deals with the oscillations of one variable between two limits. In Chapter 6, I shall review the predator–prey model, which as its name suggests deals with at least two participants that feed on one another.

The dynamics of certain events come to rest at a fixed point. In contrast there are dynamical situations where the system does not come to rest, but moves about in some manner among prescribed limits. If we equip our pendulum with a winding mechanism, thus making a grandfather clock out of it, the pendulum would oscillate between the same two extrema. Of course, it would only do so as long as the clock is wound up, or the battery of a quartz clock is charged. When neither prevails, the pendulum will again come to rest at the fixed point.

When the pendulum is pulled farther away from its operating range, it will drift back and settle to its prescribed range. As long as you do not change the length of the pendulum bob, no matter how far you pull it away from its vertical position it will assume the frequency corresponding to that length. Those who practice playing a musical instrument and who use a metronome are well aware that the position of the weight determines the metronome's tempo. Some person might inquire whether because, both in the case of the clock pendulum and the metronome, the oscillations are sensitive to the initial setting of the weight the system might be chaotic. Of course, if we change the position of the pendulum bob, or the metronome weight, the frequency will change, but this does not violate our premise. If we change the positions of the weight or the bob, we also change the physical configuration of the entire system, i.e., we now have a new stable system.

Another example of limit cycles is the human heart. We know that when we exercise, our heartbeat rises, but when we stop, it slows down to its normal beat. Going one step further, a heart that manifests erratic oscillations can be controlled with an external pacemaker. Such phenomena—whereby an oscillator gives up its own oscillations and assumes the oscillations of an applied oscillator—are called *entrainment* (Jackson[2]). Phenomena of this type also occur in plasmas when the ionized gases are confined by means of magnetic fields.

van der Pol Equation

The person who initiated modern experimental dynamics in the laboratory was Dutch electrical engineer Balthazar van der Pol[3] (1889–1959). Two nonlinear equations are named after van der Pol. The simple form without a force term follows:

$$\frac{d^2x}{dt^2} - \varepsilon\,(1 - x^2)\,\frac{dx}{dt} + x = 0 \tag{3}$$

The more elaborate version of the van der Pol equation includes a periodic force. In this case, one writes:

$$\frac{d^2x}{dt^2} - \varepsilon\,(1 - x^2)\,\frac{dx}{dt} + \omega_0^2\,x = F\cos\Omega t \tag{4}$$

It is when the frequency of the periodic force is close to the frequency of the limit cycle that the resulting periodic motion becomes entrained at the driving frequency (Moon[4]). This is why heart pacers are useful, and why their frequency must be kept in adjustment.

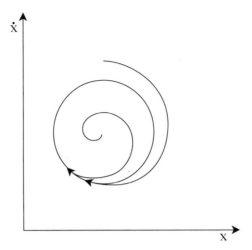

Figure 5. Limit cycle for very small {ε}.

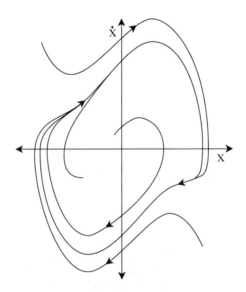

Figure 6. Limit cycle for large {ε}.

The term $\{\varepsilon\}$ is a positive constant. The variable $\{x\}$ may represent different factors because the van der Pol equation has wide applicability such as in harmonic oscillation of mechanical systems, in electrical circuits, as well as in cardiac rhythms. When $x^2 > 1$, we have positive damping, whereas when $x^2 < 1$, we have negative damping. All the trajectories located inside the limit cycle tend to move asymptotically to it, just as the trajectories on the outside drift inward to the limit cycle. The shape of the limit cycle depends on the value of $\{\varepsilon\}$. When this is very small, i.e., $\varepsilon \ll 1$ one gets trajectories similar to those shown in Fig. 5.

As can be noted, both the inner and the outer phase trajectories move towards the limit cycle. The former spirals outward towards the circle; the latter spirals inward towards the circle. When $\varepsilon > 1$, the limit cycle will look more like a rectangular hysteresis curve, which is shown in Fig. 6.

In general, the limit cycle will get elongated and assume sharp corners as the value of $\{\varepsilon\}$ increases (Davis[5]).

Duffing's Equation

Another important nonlinear differential equation is the Duffing[6] equation, which goes back to 1918. While the van der Pol equation has its roots in electrical oscillations, the Duffing equation finds many applications in mechanical vibrations. Under certain conditions it too exhibits limit cycle characteristics. It should be noted that there are different versions of the Duffing equation, e.g., without and with a forcing term, and that it has been the subject of far-reaching extensions (Moon and Holmes,[7] Ueda[8]). Here I shall present it in the forms consistent with the two forms of the van der Pol equation, i.e., both without and with a force term. In the first case the Duffing equation is commonly written as follows:

$$\frac{d^2x}{dt^2} + \omega^2 x + \beta x^3 = 0$$

(5)

With the force term it can be written as follows:

$$\frac{d^2x}{dt^2} + \omega^2 x + \beta x^3 = F \cos \Omega t$$

(6)

In these equations, $\{\beta\}$ is indicative of the hardness of the spring. Thus if $\beta > 0$ we speak of a hard spring, while if $\beta < 0$ we have a soft spring.

The next two figures show the phase plane diagrams for the Duffing equation without the force term for two different cases. In Fig. 7 may be

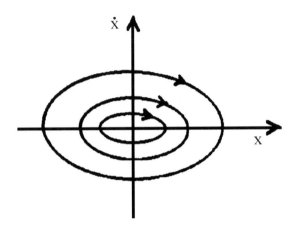

Figure 7. Duffing equation, hard spring case.

seen the situation for the hard spring case, namely, $\beta > 0$ with the concentric closed-phase trajectories.

Figure 8 shows the trajectories for the soft spring case, namely, when $\beta < 0$.

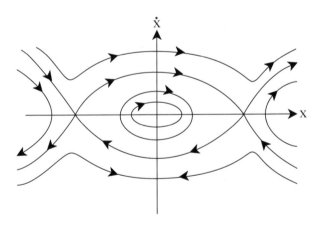

Figure 8. Duffing equation, soft spring case.

TORUS ATTRACTORS

The torus attractor is useful for systems having many degrees of freedom. One way of imagining this is to conceive of two limit cycles—Predator–Prey System-1 and Predator–Prey System-2—that are coupled, with one larger than the other. The combination appears as a torus that looks like a donut shaped sculture made of wound wire. This orbit does not repeat itself but eventually covers the entire surface of torus. The concept is similar to making a ball of yarn, although the shape is not spherical, but is ring like. Like a strange attractor the torus attractor can fluctuate in an irregular manner, but because it is energy-conserving it is predictable. In contrast, strange attractors are not information- (entropy) conserving and hence are unpredictable. The curious reader will gain a deeper understanding by perusing the enjoyable geometric presentations by R. H. Abraham and C. D. Shaw [9] of the University of California—Santa Cruz.

STRANGE ATTRACTORS

The fourth type of attractor, called the *strange attractor*, is crucial to dissipative dynamical systems that are aperiodic. Whereas the fixed-point attractor and the limit cycle manifest regularity, strange attractors tend to appear highly irregular. Closely related are *chaotic attractors*. I must admit that explaining these here will be challenging, to say the least. One reason is that they are abstract concepts that we do not recognize from experience unless we have fraternized with them on the computer monitor. The second reason for the difficulty is that to describe strange attractors properly there should be appropriate metrics. Because of their peculiar configurations this cannot be accomplished in conventional ways of measurement. We shall deal with this matter later in this chapter, as well as in later chapters after discussing the modern interpretation of entropy, information, and fractal geometry. In the meantime the reader will find the mathematical details of strange attractors in the opus by Colin Sparrow [10] of the University of California—Berkeley.

The different attractors are related, in that each demonstrates some set of pattern and regularity, even strange attractors, which are associated with chaos. This characteristic is helpful in gaining insights into phenomena that appear to be random, but nevertheless have some sort of pattern. For example, the Sierpinski gasket that we shall view in the chapter on fractal geometry is highly patterned and can be constructed geometrically. However, as demonstrated by Professor Michael Barnsley of the Georgia Institute of Technology in the PBS program "NOVA," it is also generated by placing random numbers from the throwing of a die within a triangular shape. It is evident that we have a typical case where determinism and chance meet.

The term "strange attractor" is due to D. Ruelle and F. Takens,[11] who defined it as being locally the product of a Cantor set and a two-dimensional manifold. Such attractors have complicated geometrical properties. For example, they are nowhere differentiable and have noninteger or fractal dimensions. The trajectory of strange attractors does not close on itself. They are folded, stretched, layered and undergo all sorts of contortions. The trajectories of chaotic attractors diverge, and they are sensitive to initial conditions. This is characteristic of strange attractors and of chaos. The terms "strange attractor," "fractal attractor," and "chaotic attractor" are sometimes used interchangeably. In some instances this may be permissible, but it is not always correct. Strange attractors are not necessarily chaotic. In proving this assertion, Grebogi, Ott, Pelikan, and Yorke[12] have explained that "strange" refers to the geometry of the attractor, while "chaotic" refers to the dynamics of the attractor. It is not unusual that chaotic attractors also happen to be strange. One of the best known in this category is the Hénon[13] map, which is shown in Fig. 9.

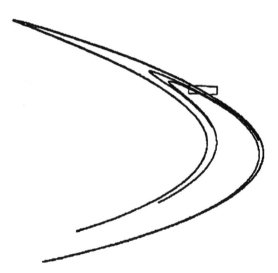

Figure 9. Hénon map. (Image generated with W. M. Schaffer's program.)

Figure 9 depicts the image of the Hénon map, which is given by the following equations:

$$x_{i+1} = ax_i(1 - x_i) + y_i \qquad (7)$$

$$y_{i+1} = bx_i \qquad (8)$$

The Hénon map is discrete in time, and its two dimensions are coupled. Figure 10 is a magnification of the boxed region in Fig. 9.

As is evident, the magnification shows much more detail. The images in both of the above figures were generated by using the software program developed by evolutionary biologist–ecologist Professor W. M. Schaffer[14] of the University of Arizona.

In contrast, we have the set of three-dimensional differential equations of Professor O. Rössler[15] of the University of Tübingen. These are:

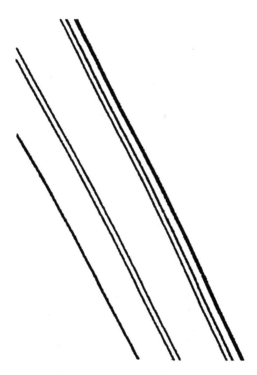

Figure 10. Magnified portion of Hénon map. (Image generated with W. M. Schaffer's program.)

$$\frac{dx}{dt} = -y - z \tag{9}$$

$$\frac{dy}{dt} = x + ay \tag{10}$$

$$\frac{dz}{dt} = xz - cz + b \tag{11}$$

where x, y, and z are the directions, and a, b, and c are parameters.

Figures 11 and 12 illustrate the Rössler attractor and were generated with the software program of W. Schaffer, G. L. Truty, and S. Fulmer.[16]

Figure 11 shows the Rössler attractor in three-dimensional space. Depending on how long one lets the program run, there will be more or less crowding of the trajectories.

Figure 12 shows the so-called Poincaré section or map. This powerful tool is obtained by slicing through the trajectories in phase space with a surface having a dimension one less than the phase space. Thus, points divide lines, lines divide surfaces, and so on. Sections can be obtained in the other planes as well.

Last but not least, we consider the Lorenz equations, which are continuous and three-dimensional. They were explored by the celebrated M.I.T. meteorologist Professor E. N. Lorenz[17] as early as 1963, but unfortunately his paper went unnoticed until almost 10 years later. The motivation behind these equations was weather prediction. What Lorenz actually found was that no long-term predictions are possible because of the

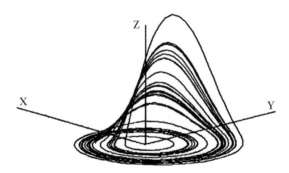

Figure 11. Rössler attractor. (Image generated with program by W. M. Schaffer, G. L. Truty, and S. Fulmer.)

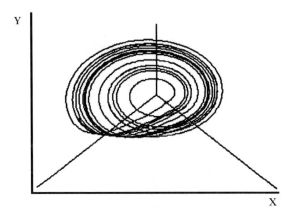

Figure 12. Poincaré section of Rössler attractor. (Image generated with program by W. M. Schaffer, G. L. Truty, and S. Fulmer.)

sensitivity to initial conditions as minute discrepancies pile up. The Lorenz equations are

$$\frac{dx}{dt} = \mathrm{Pr}\,(\,y - x)$$

(12)

$$\frac{dy}{dt} = \mathrm{Ra}/\,\mathrm{Ra}_{cr}\,x - y - xz$$

(13)

$$\frac{dz}{dt} = -\frac{8}{3}z + xy$$

(14)

In these equations, {Pr} is the Prandtl number, and {Ra} is the Rayleigh number.

An image of the Lorenz attractor generated with the computer program by Schaffer, Truty, and Fulmer[18] may be seen in Fig. 13.

An important point to appreciate is that with a fixed attractor one knows where the system will eventually find itself, but this is not so with strange attractors. The only indication one has is that the system will be somewhere on the strange attractor, but one does not know exactly where. The principal reason for this is that the dynamics of chaotic attractors never repeats itself. Nor does the trajectory of a strange attractor close on itself.

Figure 13. Lorenz attractor. (Image generated with program by W. M. Schaffer, G. L. Truty, and S. Fulmer.)

BERNOULLI SHIFT

While Newton's laws of motion are deterministic, trajectories may be irregular and contain randomness. A method that is useful in analysis is the *Bernoulli shift*, also called "baker's transformation," the "salt water taffy analog," or "Arnold's cat." I have a sweet tooth and I like chocolate and vanilla marbled fudge. In the absence of James Thurber's delightful discourse on fudge making, let us describe this with the help of Fig. 14, which is meant to constitute an area-preserving map. That is the area that remains the same in spite of all of the cutting and pasting. The image in part (a) is supposed to signify the side view of a deformable substance, in this case one layer of chocolate at the bottom, and one layer of vanilla on top. In part (b) the fudge is stretched to twice its original length, and thus its original thickness is halved. It is then cut in half and the two halves laid one on top of the other, as shown in part (c). This is then again stretched to twice its previous length until its thickness is halved (part (d)). Again it is cut and the two halves are laid one on top of the other (part (e)). If we proceed in this manner, after 2 stages there will be 4 layers, after 5 stages, 32 layers;

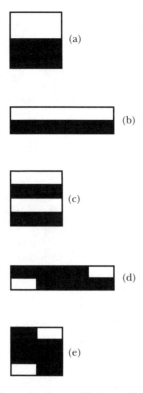

Figure 14. Layered fudge making.

and after 10 stages, 1,024 layers; because we have 2^s where $\{s\}$ is the number of stages.

It is worthwhile asking if the process can be reversed. The answer is yes. Theoretically we could pull the fudge vertically, halve it and place the two pieces horizontally side by side. Actually this optimistic statement is not strictly correct for dissipative systems. Those who are curious might try the experiment, but be forewarned that the number of stages would have to be very large. How a picture of Poincaré was scrambled and unscrambled is described by the pioneers of chaos, Crutchfield, Farmer, Packard, and Shaw,[19] who showed that the original can appear several times during the transformation process. This is called the *Poincaré recurrence*. The previous stretching and folding is typical of iterative processes that are associated with nonlinear problems. We shall visit with Arnold's cat again when we discuss the Verhulst equation and chaos.

LYAPUNOV EXPONENTIAL COEFFICIENT

The stretching and folding processes that we have observed are typical of strange attractors that depict the geometry of the dynamics. Further, the initial condition is associated with potential chaos. Let us assume that the original length of the fudge is denoted by $\{D_0\}$ and its length after exponential stretching is $\{D'\}$. We can then write

$$D' = De^{\lambda(t' - t)} \qquad (15)$$

where $\{\lambda\}$ is the *Lyapunov exponential coefficient* (LEC). The LEC measures the exponential separation of trajectories with time in phase space. These trajectories are adjacent but have slightly different initial conditions. To further conceptualize this, one envisions how a tiny sphere in phase space, having an initial dimension of $\{D_0\}$, is transformed into an ellipsoid with dimension $\{D'\}$. The reason why the LEC is indicative of chaotic conditions may be explained on the basis that nearby points separate exponentially, i.e., they separate rapidly, which suggests instability. The Lyapunov exponent can be used with conservative as well as dissipative systems. Generally, a positive LEC is taken as a strong indication of chaotic conditions. The LEC is named after Russian mathematician A. M. Lyapunov (1857–1918), who was a pioneer in developing stability analysis in nonlinear systems.

We know that fixed-point attractors have a dimension of zero (a point has no dimensions), while limit cycles have a dimension of two because they constitute a surface. Strange attractors have complex shapes, and these tend to have noninteger dimensions, also called fractal dimensions. We shall discuss this subject later. However, I do want to make the introduction to Lyapunov exponents a little clearer. In doing so I rely on the comprehensive paper by Farmer, Ott, and Yorke.[20] It is well to accept that a number of technical matters relating to the Lyapunov dimensions are not standardized; indeed, even the name can be spelled differently, i.e., Lyapunov or Liapunov.[21] In addition to Lyapunov exponents, we must also reckon with *Lyapunov numbers* and *Lyapunov dimensions*. We can dispense with explaining the Lyapunov numbers easily; the Lyapunov exponents are the logarithms of the Lyapunov numbers. While it may be relatively easy to visualize the Lyapunov exponents, it is easier to compute the Lyapunov numbers.

Still another dimension is named after Lyapunov. This is the Lyapunov dimension, $\{d_L\}$, which is defined by Farmer, Ott, and Yorke as follows:

$$d_L = k + \frac{\log(\lambda_1\lambda_2 \ldots \lambda_k)}{\log(1/\lambda_{k+1})} \qquad (16)$$

Here $\{k\}$ is the largest value for $\lambda_1 \lambda_2 \ldots \lambda_k \geq 1$. For a two-dimensional map, Eq. 14 reduces to

$$d_{\mathrm{L}} = 1 + \frac{\log \lambda_1}{\log(1/\lambda_2)} \tag{17}$$

or

$$d_{\mathrm{L}} = 1 - \frac{\lambda_1}{\lambda_2} \tag{18}$$

The Lyapunov dimension is important because it is indicative of the dynamical properties of attractors.

The Lyapunov exponent can be used with conservative as well as dissipative systems, whereas its close relative—the Kolmogorov entropy—is applicable to dissipative systems only. A positive LEC is an indication of chaotic conditions. Determining the numerical magnitude of the Lyapunov exponent is a laborious process and cannot be squeezed into this small volume. However, the procedure has been well described by Wolf, Swift, Swinney, and Vastano.[22] The LEC is a direct consequence of the pulling and stretching that we observed, and it turns out to be a very potent indicator of chaos.

BELOUSOV–ZHABOTINSKY REACTION

It is customary to express chemical reactions in the form of a stoichiometric equation as if they occurred in a single step. Actually chemical reactions occur by virtue of a set of elementary steps, and the stoichiometric equation is really a representation of the final outcome. It is also assumed generally that the reaction will proceed in a steady manner until the reactants are exhausted or an equilibrium state is reached. Mathematical and laboratory investigations have shown that these shortcuts are not necessarily warranted, and it has been observed that even apparently simple reactions can constitute complex chemical phenomena. One of the most exciting of these is oscillatory reactions. In these the products in the elementary steps rise and fall in an oscillatory manner, and they can manifest limit cycle behavior. One early experimental observation of an oscillatory chemical reaction was made by W. C. Bray[23] of the University of California, who, while studying the catalytic decomposition of hydrogen peroxide by iodate, observed that both the rate of oxygen production and the concentration of the iodine changed periodically. At about the same time, A. Lotka[24] (1880–1949) of Johns Hopkins University described oscillatory phenomena on purely mathematical grounds. He further asserted that his conclusions were ". . . the analytical confirmation and extension of an inference drawn by (the philosopher) Herbert Spencer on qualitative grounds."

The modern research on oscillating combustion as it is presently conducted finds its roots in the 1951 experimental work by B. P. Belousov, and the 1964 experiments of A. M. Zhabotinsky, and goes under the name of the Belousov–Zhabotinsky reaction, or more briefly the BZ reaction. This reaction has had a tortured history and was not appreciated immediately. My conjecture is that the existing body of knowledge did not encompass nonlinear and nonequilibrium dynamical processes. The sequence of the historical events is recounted by respected biochemist Arthur Winfree,[25] of the University of Arizona.

Belousov had noticed that when citric and sulfuric acids are dissolved in water with potassium bromate and a cerium reagent the color changes periodically from colorless to pale yellow. In his own work Zhabotinsky used an iron reagent and achieved color changes between red and blue. When the experiment is conducted in a test tube, bands appear, whereas in a shallow petri dish beautiful patterns appear and disappear.

The BZ reaction demonstrates autocatalysis, which is a form of positive feedback. In such processes the rate of the reaction depends on the concentration of an intermediary product. As the product is produced, the reaction accelerates. Thus the concentration of the species varies and it is this that gives rise to the oscillations. For an introduction to the chemical kinetics of oscillatory reactions see the physical chemistry book by P. W. Atkins[26] of the University of Oxford.

Today oscillatory reactions are generally studied in open systems, and the reactants may be replenished. This can be achieved in a continuously stirred tank reactor (CSTR). Although much more research needs to be done, Prof. Epstein and his associates[27] have suggested three conditions that are necessary for oscillatory chemical reactions to occur. These are: (a) the system must be far from equilibrium; (b) there must be feedback, in the sense that some product in the elementary steps must influence its own rate of formation; and (c) the system must exhibit bistability.

The BZ reaction has become the subject of numerous investigations. For example, Escher and Ross[28] of Stanford University have studied the limitations of traditional engines and have demonstrated how an engine powered by oscillatory chemical reactions could have greater power output due to increased energy transduction, less dissipation, and hence a higher efficiency. Since then Professor Ross has studied oscillatory reactions in a variety of situations including different reagents and laboratory setups with chemical and biological constituents.

One of the most exciting areas of study is biological reactions, which are discussed by A. Babloyantz[29] of the Brusselles School. She has described how rhythmic behavior gives rise to morphogenesis, namely, the structural development of organisms. The patterns that zebras and leopards exhibit are due to reactions that are periodic in space. Such phenomena are described with illustrations by Atkins.[30] Another related area is circadian rhythms that plants and animals experience. Such biological clocks (Winfree[31]) are receiving considerable interest in the practice of medicine in connection with the timing of surgical procedures and the administration of medication.

NOTES AND REFERENCES

1. Thompson, J. M. T., and Stewart, H. B. (1986). *Nonlinear Dynamics and Chaos*, Chichester, England: John Wiley & Sons.
2. Jackson, E. A. (1989). *Perspectives of Nonlinear Dynamics*, Cambridge: Cambridge Univ. Press.
3. van der Pol, B., and van der Mark, J. (1927). "Frequency Multiplication," *Nature*, 120(3019), 363–364.
4. Moon, F. C. (1987). *Chaotic Vibrations*, New York: John Wiley & Sons.
5. Davis, H. T. (1962). *Introduction to Nonlinear Differential and Integral Equations*, New York: Dover Publications, Inc.
6. Duffing, G. (1918). *Erzwungene Schwingungen bei Veränderlicher Eigenfrequenz*, Braunschweig, Germany: F. Vieweg und Sohn.
7. Moon, F. C., and Holmes, P. J. (1979). "A Magnetoelastic Strange Attractor," *Journal of Sound and Vibrations*, 69(2), 339.
8. Ueda, Y. (1981). "Explosion of Strange Attractors Exhibited by Duffing's Equation," in *Nonlinear Dynamics*, R. H. G. Helleman, ed. pp. 422–434, New York: New York Academy of Sciences.
9. Abraham, R. H., and Shaw, C. D. (1985). *Dynamics—The Geometry of Behavior*, Volumes 1-4, Santa Cruz, CA: Aerial Press.
10. Sparrow, C. (1982). *The Lorenz Equations: Bifurcations, Chaos, and Strange Attractors*, New York: Springer-Verlag.
11. Ruelle, D., and Takens, F. (1971). "On the Nature of Turbulence," *Communications in Mathematical Physics*, 20, 167–192.
12. Grebogi, C., Ott, E., Pelikan, S., and Yorke, J. A. (1984). "Strange Attractors That Are Not Chaotic," *Physica*, 13D, 261–268.
13. Hénon, M. (1976). "A Two-Dimensional Mapping with a Strange Attractor," *Communications in Mathematical Physics*, 50, 69–78.
14. Schaffer, W. M. (1989). *Chaos in the Classroom: I—Maps and Bifurcations*, Leesburg, VA: Campus Technology.

15. Rössler, O. E. (1976). "An Equation for Continuous Chaos," *Physics Letters*, **35a**(5), 397–398.
16. Schaffer, W. M., Truty, G. L., and Fulmer, S. (1988). *Dynamical Software*, Leesburg, VA: Campus Technology.
17. Lorenz, E. N. (1963). "Deterministic Nonperiodic Flow," *Journal of Atmospheric Sciences*, **357**, 130–141.
18. Schaffer, W. M., Truty, G. L., and Fulmer, S. (1988). *op. cit.*
19. Crutchfield, J. P., Farmer, J. D., Packard, N. H., and Shaw, R. S. (1986). "Chaos," *Scientific American*, **255**, 46–57.
20. Farmer, D., Ott, E., and Yorke, J. A. (1983). "The Dynamics of Chaotic Attractors," *Physica*, **7D**, 153–180.
21. This grammatical idiosyncracy was pointed out to me by my daughter, Emel Kopecky.
22. Wolf, A., Swift, J. B., Swinney, H. L., and Vastano, J. A. (1985). "Determining Lyapunov Exponents from a Time Series," *Physica*, **16D**, 285–317.
23. Bray, W. C. (1921). "A Periodic Reaction in Homogeneous Solution and its Relation to Catalysis," *Journal of the American Chemical Society*, **43**, 1262-1267.
24. Lotka, A. J. (1956). *Elements of Mathematical Biology*, New York: Dover Publications, Inc.
25. Winfree, A. T. (1984). "The Prehistory of the Belousov-Zhabotinsky Oscillator," *Journal of Chemical Education*, **61**(8), 661–663.
26. Atkins, P. W. (1986). *Physical Chemistry*, New York: W. H. Freeman and Co..
27. Epstein, I. R., Kustin, K., De Kepper, P., and Orbán, M. (1983). "Oscillating Chemical Reactions," *Scientific American*, **248**(3), 112–123.
28. Escher, C., and Ross, J. (1985). "Reduction of Dissipation in a Thermal Engine by Means of Periodic Changes of External Constraints," *Journal of Chemical Physics*, **82**(3), 2453–2456.
29. Babloyantz, A. (1986). *Molecules, Dynamics, and Life*, New York: John Wiley & Sons.
30. Atkins, P. W. (1991). *Atoms, Electrons, and Change*, New York: W. H. Freeman and Co..
31. Winfree, A. T. (1987). *The Timing of Biological Clocks*, New York: W. H. Freeman and Co.

CHAPTER **5**

RAPID GROWTH

INTRODUCTION

We are preoccupied with growth. Some trumpet its virtues, others deplore
its profligacy. In essence the question is not whether growth is good or bad,
but rather what kind of growth is taking place and at what rate. Because
growth and complexity are related, it is appropriate to ask the question:
"What kind of potentially impending complexity are we up against?" In
this chapter I shall discuss two early attempts to answer this question:
(a) the geometric sequence propounded by Malthus, and (b) the Fibonacci
sequence. Both are nonlinear, but the rates of growth that they model are
different. The exponential equation of Malthus is the precursor of
Verhulst's logistic equation, which in discrete form is one of the major
models used in chaos theory. We shall discuss it in later chapters. The
Fibonacci series leads to scaling, which is related to fractals. We saw in
the previous chapter that strange attractors are fractal. We shall take up
fractals in a later chapter when we learn how to determine the dimensions
of irregular shapes.

MALTHUS'S THEORY AND THE EXPONENTIAL EQUATION

In 1798 a young English clergyman by the name of Robert Thomas Malthus (1766–1834) wrote his epochal "Essay on the Principle of Population" and thereby pioneered population dynamics and political economy. The cardinal point that Malthus made[1] was that the population grows geometrically, whereas the necessary resources to sustain it increase only arithmetically, i.e., linearly. (Please see Fig. 1.) Malthus reasoned that with inadequate resources there would be poverty and misery, and the demand of the dole on the financial coffers of the British empire would increase detrimentally.

To elucidate, a progression is a sequence of numbers having a definite relation between successive quantities. For example, the series of numbers 1, 2, 3, 4, 5, 6, ... represents a linear or arithmetic progression. It is evident that the difference between successive numbers is 1, i.e., $N_{n+1} = N_n + 1$. The subscript $\{n+1\}$ means that it is the next one after the nth step. In contrast, the series of numbers 1, 2, 4, 8, 16, 32, 64, 128, ... is a geometric sequence, i.e., each succeeding number is double the previous one. It can be expressed by the relation $N_{n+1} = Nr^n$ where $r = N_{n+1}/N_n$, and $\{N\}$ is the first number in the sequence. For this geometric sequence, $N = 1$, $r = 2$, and

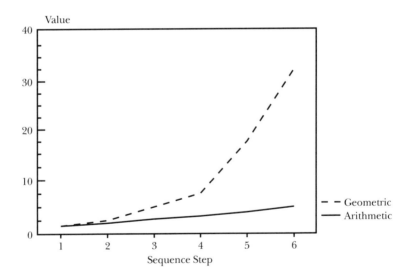

Figure 1. Arithmetic and geometric progressions.

hence the third term is 4. As may be seen in Fig. 1, changes that occur geometrically take place much faster than those that change linearly.

While the world population has increased tremendously, Malthus did not foresee the capabilities and accomplishments of modern technology. Thanks to the green revolution, the productivity of the food sector has risen astonishingly. In the U.S. in 1960, one farm worker supported about 30 persons; by 1980 the corresponding figure had almost tripled, so that each farm worker was contributing to the food supply of about 85 persons. It is quite likely that genetic engineering as it is applied to plants will increase this number even more.

Malthus's theory, also called the *exponential growth equation*, expresses the growth of the population relative to its size. It is commonly written in the form of the following differential equation

$$dN/dt = rN \qquad (1)$$

In this equation, $\{N\}$ represents the population, $\{t\}$ denotes the time, and $\{r\}$ stands for the net per capita growth rate, namely, the difference between the per capita birth and death rates.

While Malthus's equation has its origin in population dynamics, it is applicable to a wide variety of growth and decay problems. In the context of this volume, the term "population" will primarily serve as a surrogate for a host of variables in diverse fields such as economics, science, and technology. For example, in Fig. 2, the exponential increase in travel speed for different modes of transportation can be seen. Similar curves can be drawn for other variables, such as the number of computers in use, the increasing speed of computers, the miles of highways, etc.

As revealing as it can be, the exponential equation is at times misinterpreted, giving rise to doom and gloom. For example, the literature abounds with prophecies as to the complete depletion of needed resources, or that certain growth will be unending. At times such projections are flawed because they do not take into consideration factors such as technological innovation, mankind's ability to adapt, or the substitutability of resources. Equally, if not more, important, most exponential curves eventually do slow down, and bend over in the form of the letter $\{S\}$, giving rise to the term *S-curves*. We shall discuss these in the next chapter.

It is obvious from Eq. 1 that the rate of growth is sensitive to the value of $\{r\}$. Depending on the situation, the value of $\{r\}$ may or may not be constant. But even when it varies over the years, it can be treated as a constant over limited periods of time. Thus, integrating Eq. 1 between the initial time $t = t_0 = 0$, and some future time $t = t$, we obtain

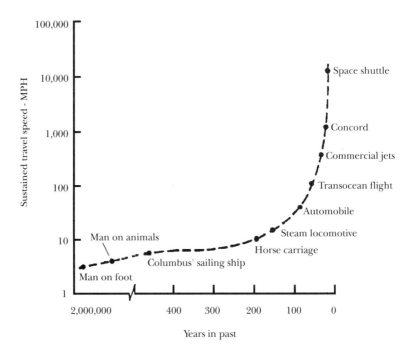

Figure 2. Vehicle velocities over the years (from N. Augustine in *Aerospace America*. Copyright (©) by American Institute of Aeronautics and Astronautics. Reproduced with permission).

$$N_t = N_0 e^{rt} \qquad (2)$$

where $\{N_t\}$ is the population (please recall that the population is a surrogate) at the future date t, $\{N_0\}$ is the population at the starting date $t = 0$, and $e = 2.71828$, namely the base of natural logarithms. An alternate, but equivalent, way of expressing Eq. 2 is to write

$$N_t = N_0 \exp(rt) \qquad (2a)$$

Both of these equations are linear in $\{N\}$, but the growth takes place exponentially (except for $r = 1$) and it is this that introduces the nonlinearity. The higher the value of $\{r\}$, the more explosive the growth will be.

From the previous equations one can calculate the accumulation of a quantity $\{N\}$ that increases at the annual net rate of $\{r\}$ after $\{n\}$ years. This is expressed as:

$$N_n = N_0 (1 + r)^n \qquad (3)$$

where $\{N_0\}$ is the original amount of the investment, $\{N_n\}$ is the accumulated amount at the end of $\{n\}$ periods, e.g., years, quarters, months, etc., including the original quantity (investment) plus the annual increases (interest).

It is not necessary for the compounding to be done at any specified period, and indeed this will vary from situation to situation. In the case of a bank loan the compounding is done essentially at the discretion of the bank. In the biological case the compounding depends on the species. It follows from Eq. 3 that the more frequently the compounding is done, the larger the final total accumulation, or the ending balance, will be. However, the final accumulation does not grow indefinitely with increasing frequency of compounding, and it eventually converges to the number 2.71828, namely $\{e\}$, the base of natural logarithms.

In the previous equations, $\{N\}$ is considered to be an integer. However, this may not be the case, hence the population is frequently treated as a smooth, continuous function in time, i.e., $\{N(t)\}$. Whether or not this is tolerable depends on the magnitude of $\{N\}$, the time interval under consideration, and the type of analysis that is to be performed. When dealing with large numbers, it does not make much difference; however, care must be exercised with small numbers.

The exponential equation can be applied widely, but there are some subtle aspects that one must be aware of under different circumstances. For example, the exponential growth of a financial investment is quite different from that of biological organisms. Hopefully, the following example will bring this out.

Example: Let us assume that we invest \$1,000 at 7% interest to be compounded annually. At the end of the first year, we would have the \$1,000 of the original principal and \$70 in interest for a total of \$1,070. At the end of the second year, the total would have gone up to \$1,144 because due to the compound interest, the 7% interest would be calculated on the \$1,070 and not on the \$1,000. In comparison, at the end of the second year, simple interest would have resulted in $1,000 + 70 + 70 = \$1,140$.

In the case of population growth, let us start with a population of one million persons and assume that the net population growth rate is 2%. At the end of the first year, the population will be 1,020,000, and at the end of the second year, $10^6 + (10^6)(0.02) + (1.020)^6 \times (0.02) = 1,040,400$, calculated quite similarly as in the case of the financial investment. Nevertheless, we must be aware of a subtle difference between the two cases. There is a flaw in what I did in order to emphasize the care that must be exercised. While every dollar invested generates interest, not every person is a reproducing person. The equations do not consider age distribution and hence fertility, habits of reproduction, and the numbers in each gender. Yet we assumed that each person in the population of 1,000,000 reproduces at the same per capita rate. It follows that the 2% annual growth rate is an average figure over a period of time and does not apply to each and every member of the society. It also includes delay because the newborn have not reached puberty and hence cannot procreate. Similarly, as people grow older they eventually become infertile and can no longer reproduce.

So far we assumed that the population inhabits a region which is a closed system so that no immigration or emigration can occur. For the case when there is immigration, Eq. 1 must be revised to take into account augmented population increase. A term must be added for this influx. Hence we write

$$dN/dt = rN + r_I N \tag{4}$$

where $\{r_I\}$ is the per capita rate of immigration, and the other symbols remain as before. If, on the other hand, there is emigration and this takes place at the per capita rate of $\{r_E\}$, then Eq. 1 must be revised to account for deceleration in the population increase. Accordingly, we write

$$dN/dt = rN - r_E \tag{5}$$

FIBONACCI SERIES

Leonardo of Piza (1175?–1250), better known as Fibonacci, grew up in Algeria and introduced Arabic mathematics to Italy. Among his contributions is the Fibonacci series of circa 1202. This series results in the Fibonacci number, f = 1.6180 . . . , which is the same ratio as the "golden section" (or "golden mean") in geometry that was applied to architecture by the Roman architect Vitruvius during the first century A.D. Once again we start with population dynamics, particularly the reproduction of hares. In later chapters we shall look at population dynamics in connection with the paired fluctuations of the hare–lynx populations, alternatively called the "logistic equation," the "predator–prey" model, or "limit cycles." Now we consider the simple breeding of rabbit couples without any predation.

In his famous volume *Liber abaci*, Fibonacci posed the following question (Kramer[2]): "If we start with a single pair of rabbits consisting of one female and one male how many rabbits will there be at the end of one year?" The following are stipulated: It takes one month for baby rabbits to become fertile, and each pair of rabbits produces one heterosexual baby couple per month. Let us determine the number of rabbit couples that can procreate each month. We start with one baby rabbit couple. At the beginning of the first month there would be that one pair only. There would also be that one couple at the end of the first month, or the beginning of the second month. At the end of the second month there would be two rabbit couples, the original couple, who are now parents, and their two offspring. In the third month there would be still be one couple that is able to breed, because it will take one month for the new pair to become productive. In the fourth month there will be two reproductive pairs, while in the fifth month there will be three, namely the original pair, their offspring, and the grandchildren. If the exercise is continued in the above fashion, there will be 5 pairs in the sixth month, 8 in the seventh month, 13 in the eighth month, 21 in the ninth month, 34 in the tenth month, 55 in the eleventh month, and 89 in the final month. This results into the following series of numbers: 1, 1, 2, 3, 5, 8, 13, 21, 34, 55, 89, ... One notes that each Fibonacci number is the sum of the previous two. It is quite evident that, unlike in a geometric sequence, the number of rabbits does not double every month. The growth is still nonlinear, but slower.

Generalizing the above, the following recursion formula emerges:

$$N_n = N_{n-1} + N_{n-2}, \text{ for } n \geq 2 \tag{6}$$

where $\{N\}$ denotes the number of rabbit couples, and the subscripts denote the generations. In this discrete dynamical system, the number at any stage depends not only on the previous one, but also on the number of earlier generations.

As before, the population serves as a surrogate in numerous applications of the Fibonacci sequence, and numbers. These include: architecture, art, biology, computers, cryptography, genetic coding, mathematics, music, network and route optimization, as well as plant phyllotaxis. For example, Fig. 3, provided by Derek Fell,[3] shows the florets of a sunflower, which follow logarithmic spirals. As may be noted, the curves have different directions. There are 55 florets in the clockwise and 89 florets in the counterclockwise spirals, both Fibonacci numbers. The observant reader will notice similar Fibonacci patterns in numerous other situations. Flowers, fruits, and vegetables are particularly rich in this regard, and it has been conjectured that this is nature's way of ensuring sustainable growth.

The Golden Mean

The infinite series of the Fibonacci numbers leads to another interesting observation. We divide successive Fibonacci numbers by one another, i.e.,

Figure 3. Photograph of a sunflower seedhead. (Copyright (©) by Derek Fell. Reproduced with permission.)

N_{n+1}/N_n, forming a continued fraction expansion. If we evaluate these ratios we find that they converge. For example, $1/1 = 1$, $2/1 = 2$, $3/2 = 1.5$, $5/3 = 1.667$, $8/5 = 1.6$, $13/8 = 1.625$, $21/13 = 1.615$, $34/21 = 1.619$, $55/34 = 1.618$, $89/55 = 1.618$, $144/89 = 1.618$. . . . Conversely, we can perform the successive divisions N_n/N_{n+1}, which approach the irrational number[4] $(\sqrt{5} - 1)/2 = 0.61803$ . . . . These values are very close to the value of the

Figure 4. Rectangle for golden mean.

celebrated *divine ratio*, or *golden mean*, $\{\phi\}$, named after Phidias of ancient Greece. In Fig. 4 may be seen how this ideal dimension is derived.

$$\phi = (a + b)/a = 1.618\ldots$$

For centuries the ratio $1.618\ldots$, or its inverse, $0.618\ldots$, has been sought after for aesthetic reasons in the human body, as well as in manmade paintings and buildings. Seashells exhibit spirals that are based on the golden section. For a rich variety of animate and inanimate objects that exhibit the golden section, see the volumes by Cook,[5] Ghyka,[6] and Huntley.[7] The golden section and the Fibonacci numbers are the inseparable partners of an enduring marriage between arithmetic and geometry.

The golden mean, $\{\phi\}$, like the base of natural logarithms $\{e\}$ and the well-known $\{\pi\}$, is an irrational number. Later we shall add another irrational number to our kit, namely, the Feigenbaum universal constant, $\{\delta\}$, which is indispensable in dealing with bifurcations leading to chaos.

NOTES AND REFERENCES

1. Malthus wrote different versions of his essay. The original, written anonymously, was polemical, and some believe that this tone was due to Malthus's disagreement with his father's altruistic views. The work offended those who believed in the perfection of man, but was welcomed by those who wished to maintain the *status quo*. As Malthus travelled and matured, he rewrote subsequent versions. Easy to locate is Malthus, T. R. (1956). "Mathematics of Population and Food," in *The World of Mathematics*, **2**, (pp. 1192–1199), Edited by J. R. Newman, New York: Simon and Schuster.
2. Kramer, E. E. (1951) *The Main Stream of Mathematics*, New York: Premier Books.
3. Derek Fell's Horticultural Library, Gardenville, PA.
4. An irrational number is one that is not the ratio of two integer numbers. An irrational number does not have a finite expansion.
5. Cook, A. T. (1979). *The Curves of Life*, New York: Dover Publications.
6. Ghyka, M. (1977). *The Geometry of Art and Life*, New York: Dover Publications.
7. Huntley, H. E. (1970). *The Divine Proportion*, New York: Dover Publications.

CHAPTER **6**

THE LOGISTIC CURVE

INTRODUCTION

Rapid growth, the focus of the previous chapter, can persist only for a limited time. The explanation is simple. If a population continues to grow geometrically, it will eventually run out of space and resources. Although we use population as a surrogate, the assertion holds for other quantities. For example, a structure can be built only so tall without losing its structural integrity and collapsing. As another example, expenditures may become so large that the budget becomes imbalanced, leading to undesirable circumstances. A more common type of growth starts gradually, accelerates, and then slows down, assuming the shape of a *sigmoid*, or S-shaped curve. In Fig. 1, the exponential and sigmoid curves for a growth starting at the same point are compared.

The common name for the sigmoid is the *logistic curve*. The abscissa is the time, and the ordinate is the applicable performance parameter. The asymptote is located where saturation is reached and further growth ceases. This is the *carrying capacity*.

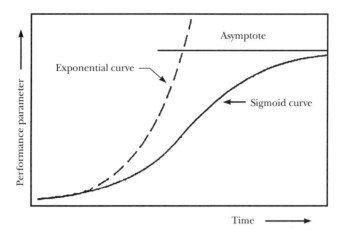

Figure 1. Comparison of exponential and sigmoid curves.

The logistic equation can assume either of two mathematical forms: (a) continuous or (b) discrete. A continuous function does not manifest sudden changes; it varies smoothly. In contrast, a set of integers is discrete and can assume widely differing values, even exhibiting jumps. Let us recall the discussion of state variables in Chapter 3. In that spirit we can represent a continuous dynamical system by the following equation:

$$\frac{dx(t)}{dt} = f[x(t)]$$
(1)

The left side of this equation is the time rate of variation of the state variable, and the right side is a description of the motion of the dynamical system. In principle, i.e., if one knew its details, Eq. 1 could serve to forecast the behavior of the dynamical system. Such computations turn out to be very ponderous; hence it is advisable to resort to computational methods. One such program, which can be used with personal computers, has been developed by University of Arizona ecologist and epidemiologist W. M. Schaffer and his associate C. W. Tidd.[1]

Frequently, $x(t)$ is multidimensional. I shall discuss this later in this chapter. In any case, the continuous form of the equation $x(t)$ means that it is defined at all times. In contrast, in the discrete form of the logistic equation, which we shall consider in the next chapter, the variables are described at only specific intervals of time. Thus, one writes the following type of equation:

$$x_{n+1} = f(x_n) \tag{2}$$

In the above equation the intervals $\{n\}$ must be integer-valued. We shall discuss this kind of equation at some length in the next chapter. It is one of the most fundamental equations of chaos theory.

The classical approach to the continuous form of the logistic equation is to deal with population numbers because the analysis was initiated by researchers in population dynamics. Admittedly, the population number $\{N\}$ is integer-valued, but for our present purposes we shall consider it to be a smooth function. This is permissible as long as we deal with large populations.

Obviously the sigmoid in Fig. 1 is a nonlinear curve. It has two equilibrium values, one at the origin, the other at the asymptote. The growth dynamics are away from the origin and towards the asymptote. It can be concluded that the origin is an *unstable equilibrium*, while the asymptote is stable, i.e., we have asymptotic stability. Further, an inflection point that is located halfway between the origin and the asymptote can be seen. Growth will accelerate up to this inflection point, and thereafter decelerate. Although the sigmoid in Fig. 1 is symmetric, this is not always the case, and the specific shape depends on the situation. The wide applicability of the logistic curve can be seen in Fig. 2, developed by engineering executive and perceptive writer Norman Augustine.[2] He has shown the logistic growth of two entirely different items.

As may be seen, both the ordinates and the abscissa have different units. Indeed it is not necessary that the abscissa be time; it could be another pertinent variable. For example, one could have a spatial series where the variation of the performance parameter is dependent on space coordinates.

In studying the growth of complex systems it is unwise to rely on a single performance parameter, because there are likely to be numerous degrees of freedom involved, and each is bound to affect the dynamics. However, even a single variable, if well chosen, can be generous with the information it reveals. Accordingly, we shall start with a single variable to understand the basic methodology, and once having mastered this, we shall examine multivariable problems.

Because the logistic curve has wide applicability, additional names for it are encountered in the literature. For example, the *Gompertz curve* is most commonly used in business applications, and the *Pearl–Reed* curve in the biological sciences. The *Lotka–Volterra* equations are generally associated with the "predator–prey" model, which we shall take up shortly. Still another name is *Verhulst's equation,* which we discuss next.

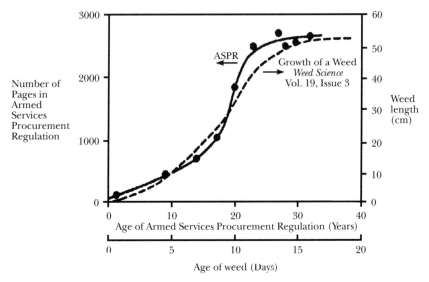

Figure 2. Logistic growth of procurement volume and height of weed. (From *Augustine's Laws*. Copyright (©) N. R. Augustine. Reprinted with permission.)

VERHULST'S EQUATION

In an 1844 publication Belgian cleric P. F. Verhulst questioned Malthus's exponential growth equation and argued that because of resource limitations the rate of growth of the population could not continue to be exponential. The original publication is difficult to locate; for details the reader is referred to the volume by Montroll and Badger.[3] I shall rely on a more intuitive approach to Verhulst's equation articulated by Prof. E. Beltrami of S.U.N.Y.[4]

According to Verhulst the population growth depends on both the per capita resources needed to survive and the maximum population that can be sustained. This means that Malthus's equation must be adjusted to take into account the negative feedback that depends on the carrying capacity $\{N_{max}\}$, namely the maximum population that can be sustained. The population that varies with time is denoted by $\{N(t)\}$. For a fixed carrying capacity, the rate of growth will slow as the population increases. The following equation results:

$$\frac{dN(t)}{dt} = rN(t)\left[1 - \frac{N(t)}{N_{max}}\right]$$

(3)

The nonlinearity of Eq. 3 becomes obvious if one multiplies it out, obtaining $[N(t)]^2$ within the brackets.

When the carrying capacity is very high relative to the population, i.e., $N_{max} \gg N(t)$, the ratio $N(t)/N_{max}$ will tend to approach zero and Eq. 3 will reduce to Malthus's equation. In general, $N_{max} > N(t) > 0$, while $dN(t)/dt > 0$. It follows that the tendency will be for $\{N(t)\}$ to grow asymptotically towards $\{N_{max}\}$. It can be shown after integration that the population at any time $N(t)$ is:

$$N(t) = \frac{N_{max}}{1 + be^{-rt}}$$

(4)

The result is particularly sensitive to the net growth rate, $\{r\}$. For example, in population dynamics, a new vaccine or other medical discovery can reduce the death rate, and hence $\{r\}$ will rise—thus the population also will increase. Similarly, a scientific discovery or a technological innovation can result in major economic change, with the associated change in jobs.

An alternate form of the logistic equation is the *delay logistic equation* (Schaffer and Kot[5]), which takes into consideration the reality that necessary resources may not be renewed instantaneously. It is written as follows:

$$\frac{dN(t)}{dt} = rN(t)[1 - N(t - \tau)]$$

(5)

where $\{\tau\}$ denotes the time lag.

CLUES FOR TECHNOLOGICAL R&D PLANNING

I have repeatedly pointed out that in nonlinear formulations, the population serves as a surrogate for other types of growth. In taking advantage of this experience, one may use a variety of other time-dependent variables, and we have seen this already in Fig. 2. One particularly useful application of the logistic curve is in evaluating potential advances in technology.

There is a common belief that technological innovation follows a linear sequence of basic research, applied research, development, testing, evaluation, manufacturing, packaging, marketing, and distribution (Tornatzky et al.[6]). While this may be so in some cases, the process can be quite nonlinear and unpredictable in other situations. For example, the steam

engine that ushered in the industrial revolution was due to the technical
ingenuity of James Watt and the entrepreneurship of M. Boulton, but did
not have the benefit of the science of thermodynamics. Recall that when
Boulton and Watt retired in 1800 from their business as prosperous individuals, Sadi Carnot, the founder of thermodynamics, was only four years
old! A more recent example is instant photography, invented by Edwin
Land, who was pressed by his small daughter as to why it was necessary to
wait for days while the pictures were being developed. In trying to respond
to the challenge, Land found little in the line of underlying science. Those
who wish to get better acquainted with the evolution of technology would
benefit from the book by George Basala.[7]

One important question that arises is for how long a product might be
improved through R&D, and when further investment is unlikely to have
significant technical payoff. This has been discussed comprehensively by
management consultant R. Foster.[8] Some of his observations are summarized in Fig. 3.

Here time is shown only indirectly, namely in terms of the cumulative
R&D expenditures. It can be seen that cotton was superseded by rayon,
nylon succeeded rayon, and polyester replaced nylon. The leveling of cord

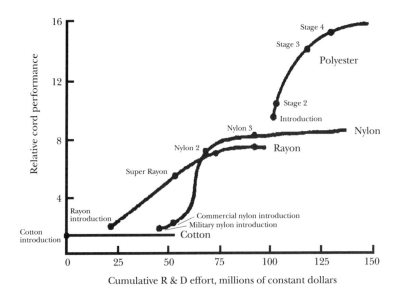

Figure 3.   Progress in cord material. (From *Innovation* by R. Foster. Copyright (©) 1986 by
McKinsey & Company. Reprinted with permission.)

performance suggests that further R&D expenditures were technically ineffective, and it might have been wiser to invest the funds on alternate R&D. When the performance of a product reaches a plateau it is best to accept the inevitable, and redirect R&D efforts. One rule of thumb is to initiate the new effort when the logistic curve such as the one in Fig. 1 reaches its inflection point. Graphs of this type can be very useful in identifying when a new tack should be taken.

Advanced technologies spawn new ones, what the computer applications pioneer J. Vallee[9] calls *relaying technologies*. Vallee has demonstrated that

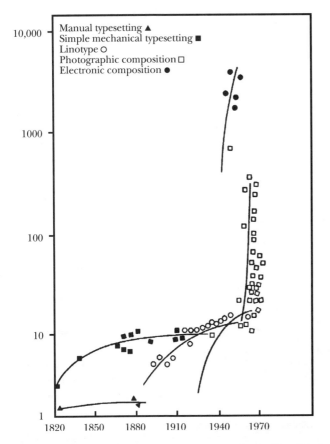

Figure 4. Advances in printed information technologies. (Reprinted from *The Network Revolution* by Jacques Vallee, by permission from And/Or Press.)

while by themselves advances in each such technologies tend to follow
S-shaped curves, the envelope of these curves when considered as a group
is not logistic, but hyperbolic. An example of this assertion is presented in
Fig. 4, where the number of characters typeset per hour (the ordinate) was
the performance parameter. It may be noted that with sophisticated printing technologies the envelope becomes almost vertical.

The envelope(s) to a series of sigmoids is useful otherwise. Planners in
developing regions can gain insights into what kind of leapfrogging is
feasible without retracing the entire history of industrial and technological
development.

The simplicity of R&D management analysis using the logistic curve,
while appealing, can be deceptive. One must be aware of at least two basic
limitations. First, the continuous logistic curve deals with a single variable,
whereas technologies have a variety of characteristics and must compete in
the market place. In the next section, we shall see how the interaction
among several competing variables can be handled. The second problem
of forecasting with the continuous logistic curve is that it does not take into
consideration chance and uncertainty. An unforeseen breakthrough can
put the sigmoid into an entirely different position and void the constraints
imposed by the limiting envelope. Events that involve chance are better
explored with the discrete form of the logistic equation.

LOTKA–VOLTERRA EQUATIONS

The dynamics of many complex systems cannot be described by the behavior of a single parameter. Accordingly, one must be prepared to deal with
interactions among several variables. Once again the basic model is popu-
lation-oriented, and it is commonly called the *predator–prey* model. Specifically, one inquires at what rates the interacting species increase or
decrease. The origin of the rate equations of competing species goes back
to the 1910 work of A. J. Lotka dealing with autocatalytic processes, and the
1928 work of V. Volterra regarding fish catches in the Adriatic Sea.

Consider a pastoral setting where rabbits feed on grass, foxes feed on
rabbits, and foxes are hunted for their pelts to be sold by furriers. The
sequence of events can get quite involved with a variety of exogenous effects.
However, because I want to present the basic model, I shall limit the
discussion to rabbits and foxes. We can write:

Rabbits + foxes ⇒ fewer rabbits and more foxes
More foxes ⇒ fewer rabbits

Fewer rabbits ⇒ fewer foxes
Fewer foxes ⇒ more rabbits, and so on.

Two observations follow: First, while the numbers of rabbits and foxes are interdependent, there would be no foxes without rabbits, whereas without foxes the rabbits would become overpopulated. This is depicted in Fig. 5, wherein I use the terms "predator" and "prey" for purposes of generality.

It is not necessary for the predator to be physically larger than the prey. A leopard can kill an elephant, a submarine can sink a huge aircraft carrier, bacteria and viral organisms too small to be seen by the naked eye can lead to the death of plants, animals, and humans. One also recalls Gulliver's travels in Lilliput, as well as the Biblical story of David and Goliath.

Second, a cyclical process is evident. If the rabbit and fox populations remain in dynamic balance, this can go on indefinitely. In Fig. 6 are shown lynx and hare data obtained for the Hudson Bay Company, which was prominent in the fur trade. The graphs are adaptations of data given by MacLulich.[10]

As is evident there is a seesaw pattern. This suggests that the predator and prey populations are coupled to one another. Hence, one would expect the limit cycle orbits that we discussed in Chapter 4, and which we see in Fig. 8 of this chapter. However, upon close examination of Fig. 6, some peculiarities

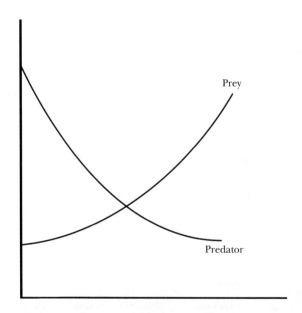

Figure 5. Overpopulation of prey, starvation of predator.

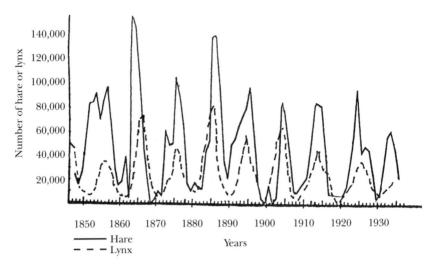

Figure 6. Lynx–hare population fluctuations. (Source. "Fluctuations in the Numbers of the Varying Hare," by D. A. MacLulich. Adapted with the permission of the University of Toronto Press.)

may be noted. This has led M. E. Gilpin[11,12] to inquire, "Do hares eat lynx?" To explore this question, corresponding lynx and hare numbers on 11 dates were read off from the population number vs. year coordinate system shown in Fig. 6. These were then replotted on a new predator vs. prey coordinate system that is shown in Fig. 7. The data points fell within a two-decade period. The reason for limiting myself to two decades was so that the graph would not be cluttered. As can be seen, the points follow a zigzag pattern reminiscent of Brownian motion. This is quite different from the concentric orbits in Fig. 8, which are normally construed as predator–prey phase space. This observation has far-reaching implications. In the real world, we use many pairs of curves, such as the leading indicators pacing the progress of one economic sector or another. Pairs of curves are also used in meteorology, such as the atmospheric temperature following the carbon dioxide content (the greenhouse effect). It must be realized that the conclusion will be quite different if the data are chaotic when plotted in the predator–prey space, as in Fig. 7, or if the data represent limit cycles, as in Fig. 8. One should not jump to conclusions.

My students and I have looked at different predator–prey type data and observed similar chaotic paths. Further, we noticed that the particular configuration of the plots in the predator–prey space can vary significantly

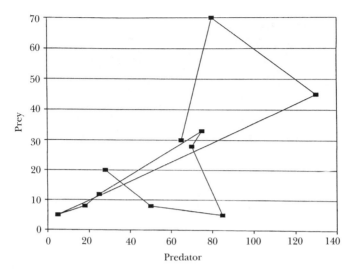

Figure 7. Predator and prey data from Fig. 6 plotted in predator–prey space using Quattro Pro 4.0 graphics software.

depending on whether the entire data set is plotted, or only short range data sets are considered. This suggests that the coupling of the predator–prey may be more than purely mechanistic and that synergistic effects may develop. Analyses of this type, if pursued in greater detail, should give clues to cause-and-effect mechanisms that had not been suspected originally.

Theoretically, if we take corresponding data points for predator and prey populations, and plot them in the predator–prey phase space, we get the closed trajectories shown in Fig. 8, generated with the computer program of Professor Ralph Abraham and L. Norskog[13] of the University of California–Santa Cruz. Here the horizontal axis represents the number of predators, and the vertical axis the number of prey. The trajectories arise from the vector field in the phase plane shown in Fig. 9, also generated with the Abraham–Norskog program.

The idealized predator–prey system constitutes a closed ecological system, and it is called *self-reproductive*, or *autocatalytic*. This and the periodicity that is evident suggest one type of limit cycle. These depict stable oscillations, such as the regular oscillations of a grandfather clock pendulum. Limit cycles are typical of nonlinear equations. Therefore, the equilibrium solutions of nonlinear equations need not be stable points, but may be stable limit cycles.

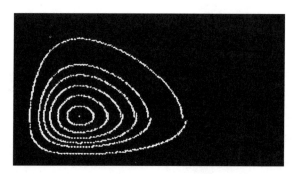

Figure 8. Trajectories of predator–prey model in phase space. The horizontal axis represents predators, the vertical axis the prey. (Generated with R. Abraham–L. Norskog's computer program.)

At the center of the closed trajectories, predators and the prey will be at their equilibrium point, also called the fixed-point attractor. However, it should be noted that the dynamics of the predator–prey system are different from the dynamics of van der Pol systems, which we shall learn about in connection with strange attractors.

There is also a delay between the two populations. For example, a shortage of hare leads to a shortage of lynx, and then the hare population starts increasing, and eventually the lynx population comes back. The delay is determined by the reproduction rate and the time it takes for the animals to come to maturity.

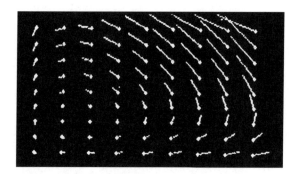

Figure 9. Vector field in predator–prey phase space. (Generated with R. Abraham–L. Norskog's software program.)

The predator–prey model is quite broad in scope and has numerous applications in a wide variety of disciplines. These include the competition between two companies or between political adversaries. In the health sciences, immunology presents an excellent example. The immune system, comprised of organs and cells, responds to intruding bacteria and viruses. In the realm of physics, a kicked rigid rotator, much like pushes given to a child's swing, is a useful paradigm. Many other examples can be cited, but the important point to appreciate is that when several factors are found in the same complex system, four basic situations can occur. The trivial one is that the different components may coexist without interacting with one another. The second possibility is competition, and the third is a subtle variant of this, namely, combat. Finally, there can be synergism, or self-organization. Any one of these may evolve into another through some sort of a stimulus that brings about a phase transition. For example, it is presently believed that not one but a number of different factors must coexist to cause a disease. This is why health professionals advise us to shy away from factors that are known to increase risk. Greater detail concerning the various models of interacting populations may be found in Peschel and Mende.[14]

The analysis proceeds as follows: Let the number of the prey be denoted by $\{N_1\}$, and the number of predators by $\{N_2\}$. The interaction between these species was expressed by Volterra in the form of a pair of rate equations:

$$\frac{dN_1}{dt} = \alpha_1 N_1 - \lambda_1 N_1 N_2 \tag{5a}$$

$$\frac{dN_2}{dt} = -\alpha_2 N_2 + \lambda_2 N_1 N_2 \tag{5b}$$

where $\{\alpha\}$ and $\{\lambda\}$ are positive interaction coefficients.

If there were only one species, i.e., if either $N_1 = 0$ or $N_2 = 0$, Eqs. 5a and b would reduce to Malthus's equation. If both species exist, and one preys on the other, then in Eq. 5a, the term $\{-\lambda_1 N_1 N_2\}$ represents the rate of decrease of the prey, while in Eq. 5b, the term $\{\lambda_2 N_1 N_2\}$ represents the rate of increase of the predator, and $\{\alpha N\}$ is the autocatalytic term. Due to the nonlinear terms they contain, Eqs. 5a and b give oscillatory solutions as one might expect from the earlier phenomenological discussion.

Equations 5a and b can be generalized to the case of n-species (Goel, Maitra, and Montroll[15]), as follows:

$$\frac{dN_i}{dt} = k_i N_i + \frac{1}{\beta} \sum_{j=1}^{n} a_{ij} N_i N_j \tag{6}$$

In this equation the term on the left side and the first term on the right side constitute Malthus's equation and $\{k_i\}$ is the rate coefficient. When $k_i > 0$ there is exponential growth, while when $k_i < 0$ there is exponential decay.

It is the second term on the right side, namely the nonlinear quadratic term, that introduces the interactive complexities. Here $\{N_i N_j\}$ is the number of possible binary encounters between species $\{i\}$ and $\{j\}$. The *competition coefficients* $\{a_{ij}\}$ describe the rate of the encounters among the species and, depending on the situation, they may be positive, negative, or zero. The latter indicates that while the species do coexist, they do not interact. Under certain circumstances this situation may be modified by appropriate brokerage if interaction is desirable.

The Lotka–Volterra equations are applied widely to a host of different problems in biology, economics, neural networks, physics, and even adversarial relations among nations. I shall briefly review two basic models formulated by Beltrami.[16] We start with what he calls the "competition model." Here two populations, {1} and {2}, vie for the same food supply, although it could just as well be two organizations that are after the same resource. Even indirect interaction will interfere with the growth of both populations. One can write:

$$\frac{dN_1}{dt} = \alpha_1 N_1 \left(1 - \frac{N_1}{N_{\max - 1}} \right) - \lambda_1 N_1 N_2 \tag{7a}$$

$$\frac{dN_2}{dt} = \alpha_2 N_2 \left(1 - \frac{N_2}{N_{\max - 2}} \right) - \lambda_2 N_1 N_2 \tag{7b}$$

In these equations $N_{\max - 1}$ and $N_{\max - 2}$ are the carrying capacities for the two populations, respectively, while the constants λ_1 and λ_2 are indicative of the relative competitiveness of the two populations.

The second model of Beltrami is the "combat model." Simply interpreted, combat is a type of competition. The frame work of this model benefits from the early work of F. W. Lanchester,[17] which deals with the dependence of various combat variables on the force strength. For example, losses suffered by one side are proportional to the size of the forces of the other side.

$$\frac{dN_1}{dt} = -\alpha N_1 - \beta N_1 N_2 \tag{8a}$$

$$\frac{dN_2}{dt} = -\gamma N_2 - \lambda N_1 N_2 \tag{8b}$$

In Eqs. 8a and b, all constants are positive. A minus sign is placed in front of each term because combat constitutes attrition.

The case of competition between three species has been formulated by May and Leonard.[18] They start with the quadratically nonlinear Gauss–Lotka–Volterra equations for $\{n\}$ competing species:

$$\frac{dN_i(t)}{dt} = r_i\, N_i(t)\left[1 - \sum_{j=1}^{\infty} \alpha_{ij}\, N_j(t)\right]$$

(9)

They invoke the symmetry assumption $r_1 = r_2 = r_3 = r$, and insist that the populations affect one another in the same manner. Thus $\alpha_{12} = \alpha_{23} = \alpha_{31} = \alpha$, and $\alpha_{21} = \alpha_{32} = \alpha_{13} = \beta$. Accordingly,

$$\frac{dN_1}{dt} = N_1[1 - N_1 - \alpha N_2 - \beta N_3]$$

(10a)

$$\frac{dN_2}{dt} = N_2[1 - \beta N_1 - N_2 - \alpha N_3]$$

(10b)

$$\frac{dN_3}{dt} = N_3[1 - \alpha N_1 - \beta N_2 - N_3]$$

(10c)

Using this approach as a pattern, one can write the equations when there are even more competing species. The problem, of course, is having reliable information for the various coefficients.

The foregoing constitutes a review of the continuous form of the logistic equation, but in no way exhausts the subject. The logistic equation can be particularly useful in dealing with problems where experience is lacking. Ideally, when confronted with a problem one tries to identify the appropriate fundamental scientific law within the purview of which the problem would be likely to fall. This is not always possible. There may be no applicable fundamental law, or there may simply not be sufficient information. In such cases it is useful to look for a suitable distribution function. For example, one might explore a Gaussian or normal distribution function and solve for two moments, or one might try the Poisson function and solve for the first moment. If these attempts do not prove to be useful, the application of the continuous logistic equation is often rewarding. However, in some cases there may be stability issues, in which case the application of the discrete form of the logistic equation can reveal insights. Indeed this approach is most helpful in opening entirely new avenues to the understanding of nonlinear dynamic problems that in the past eluded our efforts. We shall explore these in the next chapter.

NOTES AND REFERENCES

1. Schaffer, W. M., and Tidd, C. W. (1991). *NLF: Nonlinear Forecasting for Dynamical Systems,* Leesburg, VA: Campus Technology.
2. Augustine, N. R. (1986). *Augustine's Laws,* New York: Viking.
3. Montroll, E. W., and Badger, W. W. (1974). *Introduction to Quantitative Aspects of Social Phenomena,* New York: Gordon and Breach Science Publishers.
4. Beltrami, E. (1987). *Mathematics for Dynamic Modeling,* Cambridge, MA: Academic Press.
5. Schaffer, W. M., and Kot, M. (1986). "Differential Systems in Ecology and Epidemiology," in *Chaos* (pp.158–178), A. V. Holden, ed. Princeton, NJ: Princeton Univ. Press.
6. Tornatzky, L. G., Eveland, J. D., Boylan, M. G., Hetzner, W. A., Roitman, D., and Schneider, J. (1983). *Technological Innovation: Reviewing the Literature,* Washington: National Science Foundation Report.
7. Basala, G. (1988). *The Evolution of Technology,* Cambridge: Cambridge Univ. Press.
8. Foster, R. (1986). *Innovation,* New York: Summit Books.
9. Vallee, J. (1982). *The Network Revolution,* Berkeley, CA: And/Or Press, Inc.
10. MacLulich, D. A. (1937). *Fluctuations in the Numbers of the Varying Hare-(Lepus Americanus).* Toronto: Univ. of Toronto Press.
11. Gilpin, M. E. (1973). "Do Hares Eat Lynx?" *The American Naturalist,* 107(957), 727–729.
12. Gilpin, M. E. (1979). "Spiral Chaos in a Predator–Prey Model," *The American Naturalist,* 113 (2), 306–308.
13. Abraham, R., and Norskog, L. *Periodic Attractors in the Plane,* Santa Cruz, CA: Aerial Press.
14. Peschel, M., and Mende, W. (1986). *The Predator–Prey Model,* Wien, Germany: Springer-Verlag.
15. Goel, N. S., Maitra, S. C., and Montroll, E. W. (1971). "On the Volterra and Other Nonlinear Models of Interacting Populations," *Reviews of Modern Physics,* 43, 231–276.
16. Beltrami, E. (1987). *op. cit.*
17. Lanchester, F. W. (1916). *Aircraft in Warfare, the Dawn of the Fourth Arm,* London: Constable and Company, Ltd.
18. May, R. M., and Leonard, W. J. (1975). "Nonlinear Aspects of Competition between Three Species," *Siam Journal of Applied Mathematics,* 29(2), 243–253.

CHAPTER 7

THE DISCRETE LOGISTIC EQUATION

INTRODUCTION

Previously we considered the logistic equation as a continuous function of time. In this chapter we shall concentrate on the discrete form of the logistic equation. Discrete dynamical systems differ from continuous ones in that variations are not continuous, but are points. Instead of dealing with trajectories, one deals with *itineraries*, namely, sequences of discrete points, $x_1, x_2, \ldots, x_{n-1}, x_n, x_{n+1}$, (Guckenheimer and Holmes[1]). The basic form of the discrete logistic equation is commonly written as follows:

$$x_{n+1} = f(x_n) \qquad (1)$$

where the increment $\{n+1\}$ follows the increment $\{n\}$. The form of the function on the right side may differ, depending on the situation under consideration.

The distinction between the two forms of the logistic equation is more than academic and has far-reaching implications in understanding complexity and chaos theory. In general, real-life problems are not continuous, but consist of discrete elements like molecules, cells, the parts in a batch process, or intangible inputs and outputs such as bits of information.

There are a number of reasons why the discrete form of the logistic equation recommends itself. First, we know that complex problems involve chance events. This means that there can be a discontinuity, and hence a discrete equation is natural to expect. Such discontinuities need not be random, but can be periodic. For example, certain biological species like the gypsy moth reproduce at regular, discrete intervals; companies issue reports annually; and grades are posted at the end of semesters. Second, the logistic equation is useful when dealing with initial value problems,[2] one of the crucial characteristics of chaos. There is also a compelling pragmatic reason for writing difference equations, namely, the ubiquitous computer, which makes iterative computations painless. It is particularly gratifying that with available software programs many such computations can be performed even with personal computers.

The discrete logistic equation represents the continuous logistic equation stroboscopically—like snapshots taken at regular periodic intervals. Both types of logistic equations are very useful, but in many ways the discrete logistic equation constitutes the foundation of chaos theory. The continuous and the discrete logistic curves are compared, respectively, in Figs. 1 and 2.

There are some fundamental differences between the logistic curves depicted in Figs. 1 and 2. In the former we see the single value of the performance parameter at any instant in time. In Fig. 2, this single value variation still occurs, but quickly gives way to bifurcations. What this means

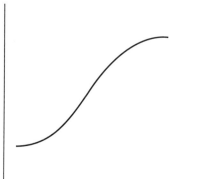

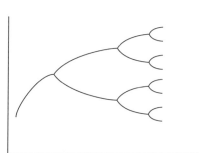

Figure 1. Continuous logistic curve.   **Figure 2.** Discrete logistic curve.

is that as we go away from equilibrium, a number of alternative stable states develop. Call them choices or options as you wish, or call them opportunities for advancing from being to becoming. The American poet Robert Frost put it beautifully in his famous poem, "The Road Not Taken":

> I shall be telling this with a sigh
> Somewhere ages and ages hence:
> Two roads diverged in a wood, and I—
> I took the one less traveled by,
> And that has made all the difference.

During complex events the divergence is not limited to two roads, but there can be many. In this chapter we shall study *bifurcation* phenomena from the viewpoint of dynamics; in general, this is quite adequate. In Chapter 3, we discussed equilibrium and stability (or their converses). In the chapter differentiating among the various types of entropy we will pay homage to the far-from-equilibrium thermodynamics promulgated by Ilya Prigogine and his Brusselles School. Specifically, we shall see that certain processes are not only dynamic, but also thermodynamic in nature. This observation opens up entirely new vistas of learning, such as self-organization.

THE DISCRETE LOGISTIC CURVE

Frequently, the discrete logistic equation is presented axiomatically. While this is certainly acceptable, we shall review its formulation to show how it relates to the continuous form. I hope that in this manner some insights will be introduced to reduce the mathematical mystery when one first sees a graph with numerous branches, blank spaces, and hashed areas all generated by the computer. Please jump ahead and check out the computer images of the discrete logistic equation. To save confusion, please note that names given to either form of the logistic are at times the same, i.e., logistic equation, Lotka–Volterra equation, and Verhulst's equation. It is important therefore to be aware which mathematical form is being used.

Figure 2 differs quantitatively as well as qualitatively from Fig. 1, with its characteristic *pitchfork bifurcations*. Not only can the variable $\{x\}$ represented along the ordinate embark on different paths, but it can also assume different values at the same time juncture. The abscissa denotes a parameter $\{a\}$ that is indicative of the strength of the nonlinearity, or the departure from equilibrium. Alternate names of this parameter are the "bifurcation," the "map," or the "control" parameter.

A number of bifurcations are evident in Fig. 2. Bifurcations are an indication of qualitative changes in the dynamics of nonlinear systems. The discrete form of the logistic equation is also called the *bifurcation map*, and with it is associated *period doubling*. Period doubling is an underlying factor in the occurrence of complexity. In Fig. 2, the first bifurcation has two branches, the second has four branches, then eight, and so on. We talk of these as period-2, period-4, and period-n doubling. More specifically, in period-4, the period length is twice that in period-2. The term "chaos" became au courant when Li and Yorke[3] published their famous paper "Period Three Implies Chaos," wherein they conjectured that chaos occurs with odd-numbered cycles or aperiodicity. Chaos is not apparent in Fig. 2, but will become quite evident in later figures in this chapter.

The general form of the continuous logistic equation was given previously and is recalled here:

$$\frac{dN(t)}{dt} = rN(t)\left[1 - \frac{N(t)}{N_{\max}}\right]$$

(2)

Its counterpart in discrete form is usually presented as follows:

$$x_{n+1} = ax_n(1 - x_n)$$

(3)

Some resemblance between these equations is evident. Nevertheless these two equations are quite different in form and in substance, and they are both useful in describing the evolution of dynamic systems, albeit different aspects. As a crude rule of thumb, it is advisable to apply the discrete form when the value of the parameter {a} is greater than 2.5.

The variable {x} is a positive number less than unity, i.e., $0 < x < 1$, because it is usually normalized with respect to the carrying capacity, i.e., $x_n = N_n/N_{\max}$. For example, {x_n} is the value of {x} at the time increment {n}, {x_{n+1}} is the value of {x} at the next time increment, namely at the time instant {$n+1$}, {x_n+2} is the value at the following instant of time, {$n+2$}, and so on. It is just like measuring some indicator at successive clicks of a digital watch. As in the case of the continuous logistic curve, the normalized population number {x} serves as a surrogate for all sorts of variables. The map parameter {a} in Eq. 3 is always positive, but may fall within either of two ranges depending on the proclivities of the user. Both are indicative of the nonlinearity of the equation and involve dissipative effects typical of real-life problems. The two ranges encountered are either $0 < a < 1$, or $0 < a < 4$. It is important in numerical computations to be aware which is being applied.

THE MORPHOLOGY OF THE DISCRETE LOGISTIC EQUATION

There are different ways of deriving the discrete logistic equation. Probably the most formal approach is to apply numerical methods of initial-value differential equations. These techniques include Taylor series, Galerkin solutions, and Runge–Kutta techniques (Carnahan, Luther, and Wilkes,[4] or Jackson[5]). The latter are popular in the study of dynamical systems because of their accurate integration. Another approach is to configure Eq. 3 by analogy from Eq. 2 (Beltrami[6]). It has been derived from a resource viewpoint by Professors S. S. Penner and D. F. Wiesenhahn,[7] of the Energy Center at the University of California–San Diego, who studied the stability of growth rates in energy technologies. Finally, an empirical approach is quite permissible.

In a pacesetting 1976 paper, Robert May[8] pointed out that simple mathematical models can represent very complicated dynamics. He began by considering how a future population is related to a previous generation, and then wrote:

$$N_{n+1} = F(N_n) \tag{4}$$

This is a first-order, nonlinear difference equation, and the basic form has been used in different fields. A variety of versions of the function $F(N_n)$ for different applications in biological literature has been compared by May and Oster.[9] The form of $F(N_n)$ that May emphasized was:

$$N_{n+1} = N_n(a - bN_n) \tag{5}$$

By letting $x = bN/a$, one can then write:

$$x_{n+1} = ax_n(1 - x_n) \tag{6}$$

If this equation is multiplied out as before, one obtains:

$$x_{n+1} = ax_n - ax_n^2 \tag{7}$$

wherein the first term on the right side is positive and represents linear growth, which will be steady if $\{a\}$ is a constant. The second term on the right is squared, and this makes it quadratic or nonlinear. The growth rate is not always constant, leading to, at times, counterintuitive results. We shall appreciate this better when we study *return maps* in a subsequent section.

When the normalized population, $\{x_n\}$, is very small, the nonlinear term can be neglected, and we get $x_{n+1} \approx ax_n$, thus indicating a substantially linear growth. It follows that when $a > 1$, the population will increase. As the value of $\{x_n\}$ rises, the second term becomes significant, and first an inflection point and then a steady state condition occur when the two terms—although equal in value—oppose one another because of their opposite signs. These events depend on the value of {a}. When $a < 1$, the population will always decrease and eventually becomes extinct regardless of the size of the initial population. A variety of events is possible when $a > 1$. These matters will be discussed when we take up the return maps. However, it is not too early to indicate that when $a \geq 4$ the solutions lose their reality. For example, the number could give negative normalized populations! On the other hand, ergodicity occurs when $a = 4$.

If the population is allowed to increase to a sufficient level, the nonlinear term {ax_n^2} will become significant and hence can no longer be neglected. Further, because there is a negative sign in front of this nonlinear term, we are confronted with a decrease in the population, which may well be due to depleting resources. This observation is consistent with Verhulst's fundamental argument. It is apparent that the discrete logistic equation does not contradict the continuous logistic equation, although it does provide us with additional insights when instabilities can be expected.

An alternate way of writing the discrete logistic equation is as follows:

$$N_{n+1} = rN_n(\text{Mod } 1) \tag{8}$$

where the statement (Mod 1) means that the integer part is being dropped.

RETURN MAPS

The numerical solution of the discrete logistic equation, i.e., Eq. 6, yields the *return map*, which is an inverted parabola. The maximum point of x_{n+1} equals $a/4$ and occurs when $x_n = 0.5$. The configuration of this map varies depending on the value of the parameter {a} and the initial population {x_0}.

It has been shown by May,[10] and Jensen,[11] that the progress of dynamic systems can be described graphically with the use of the quadratic map. In this graphical procedure one plots the normalized x_n as the abscissa, and the normalized x_{n+1} as the ordinate for different values of $a > 1$. The resulting parabola has intercepts at $x_n = 0$, and $x_n = 1$, while the value of x_{n+1} varies with the chosen value of {a}, but its maximum always occurs at $x_n = 0.5$. Using Eq. 5, one first constructs the inverted parabola for different values of {a}. Then the diagonal 45° line is drawn from (0,0) to (1,1) on the x_n vs. x_{n+1} plane. Next one picks some initial value of $0 < x_0 < 1$, and draws a vertical line to intersect the parabola. The moment the initial value x_0 is fixed, the evolution of the dynamic system is preordained for the chosen value of $\{a\}$. This is done by iteration. One simply picks an original value for x_n and calculates x_{n+1} for the given value of $\{a\}$. This value is then entered in the quadratic equation as x_n, and the next x_{n+1} is calculated. In other words, $x_1 = f(x_0)$, $x_2 = f(x_1)$, $x_3 = f(x_2)$, ... until $x_n = f(x_n - 1)$. The procedure may be seen in the imaginary iteration machine shown in Fig. 3.

The iterative approach suggests the use of numerical computations. One may either write a program in the language of one's choice, or use a spreadsheet if that is more familiar. In the latter case one can proceed as follows: 1) Label cell A1: $\{a\}$, label cell B1: $\{x_n\}$, label cell C1: $\{1 - x_n\}$, label cell D1: $\{a * x_n * (1 - x_n)\}$ and label cell E1: $\{x_{n+1}\}$. 2) Decide on an initial value of the parameter $\{a\}$, being aware whether the range will be $0 < a < 1$, or $0 < a < 4$. Treat this as a constant throughout this run. 3) Decide on the initial value of $x_n = x_0$. Run the program for different values of $\{a\}$ and $\{x_0\}$. The tables of numbers that you will get will be quite interesting. More revealing would be their graphical representation. You should be able to do this if your spreadsheet program has its own graphics option. If not, you should be able to import the data you calculated into a freestanding graphics program.

Plotting the logistic equation for different values of $\{a\}$ and $\{x_0\}$ gives revealing insights into population dynamics. Figure 4 depicts the graph when the initial value was picked to be $x_0 = 0.91$ for the logistic equation written as $x_{n+1} = 4x_n(1 - x_n)$. These were computed by my student, Blaine D'Amico.[12]

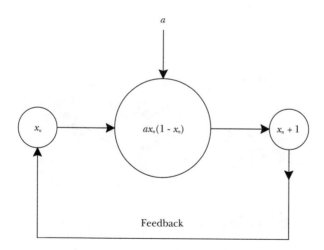

Figure 3. Imaginary iteration machine.

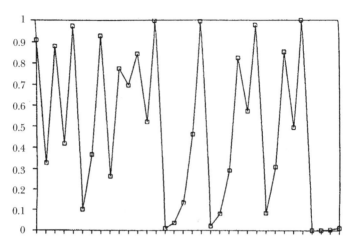

Figure 4. Plot of $x_{n+1} = 4x_n(1 - x_n)$ for $x_0 = 0.91$. (Source: B. D'Amico. Reproduced with permission.)

The quadratic equation is very sensitive to *initial conditions*. For example, if the initial condition had been chosen 0.01 less, i.e., $x_0 = 0.90$, the curves would have diverged as early as the second generation. The divergence depends on the conditions. For example, for the same equation the curves would have diverged for a difference in the initial condition of still 0.01 if $x_0 = 0.70$ in the third generation. A geometric approach simplifies the recursive calculations and can provide some profound insights. This may

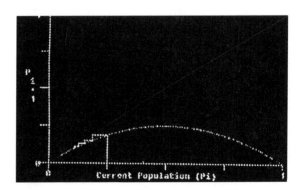

Figure 5. Return map for normalized initial population = 0.25, and $a = 0.25$. (Computed with H. Levitan's program.)

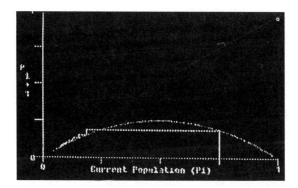

Figure 6. Normalized initial population = 0.75, and $a = 0.25$. (Computed with H. Levitan's program.)

be seen in Figs. 5 and 6, which were constructed using the software program developed by Professor H. Levitan[13] of the University of Maryland. (Please note that in Figs. 5–9, which were captured from the computer monitor screen, the normalized population is denoted by $\{P\}$ instead of $\{x\}$, as I have been using.) In the Levitan software program, the range for the parameter is $0 < a < 1$. However, in the following discussion I have used the parameter range $0 < a < 4$. I warn the reader about such potential differences. Either is acceptable.

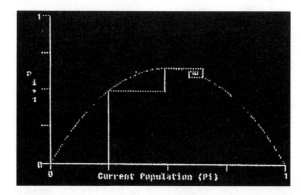

Figure 7. Initial normalized population = 0.25, and $a = 2.6$. (Computed with H. Levitan's program.)

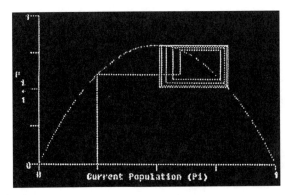

Figure 8. Initial normalized population $0 = .25$, and $a = 3.2$. (Computed with H. Levitan's program.)

It can be seen in Figs. 5 and 6 that regardless of the initial population, when $a < 1$, the population becomes extinct because x_{n+1} converges to zero. The point of intersection of the diagonal with the parabola is a fixed point on the map, or an attractor. Very small population growth rates will always result in convergence to this fixed point, and hence the population will become extinct.

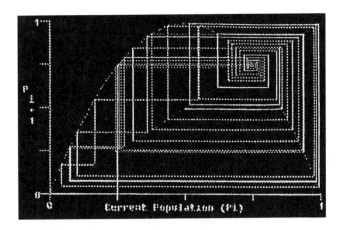

Figure 9. Initial normalized population $= 0.25$, and $a = 4$. (Computed with H. Levitan's program.)

The situation changes quite significantly when $a > 1$, as is shown in Figs. 7, 8, and 9. Even with a small initial population the population will increase, and eventually the $\{ax_n^2\}$ term will become significant. However, the population will ultimately recede because of the negative sign in front of the quadratic term. This is explained by Verhulst's argument.

When $a > 1$, the diagonal intersects the parabola at $x = (a-1)/a$, and the fixed point becomes unstable. For $1 < a < 3$, the graphical iterations converge to a fixed or stable point, i.e., a nonzero equilibrium population is achieved. However, when $a > 3$, the fixed point becomes unstable. Indeed there may be more than one fixed point, one high and one low, for x_n, and the population will oscillate between these limit cycles. As $\{a\}$ becomes larger and tends to approach 4, more and more fixed points arise, and $\{x_n\}$ wanders all over the map in a random manner, suggesting chaos.

The range of map parameter $\{a\}$ depends on how the software program is written and is a matter of taste. In the Levitan program, $0 < a < 1$, although frequently one encounters $0 < a < 4$. As indicated in the captions of Figs. 5–9, the value of $\{a\}$ was converted to the range $0 < a < 4$.

BIFURCATION DIAGRAM

A major problem in the understanding of complex dynamical systems is being able to identify when the behavior is periodic and when it might lapse into chaos. To resolve the matter one must come to grips with two issues. The first is the realization that the routes to chaos are many (Holmes[14]), and any direct causal determination is difficult. Second, as in any diagnostic procedure, there are numerous techniques that must be applied. One of the most powerful techniques is the bifurcation diagram shown in Fig. 10 for the logistic equation. Here x_{n+1} in Eq. 6 is evaluated for different values of the parameter $\{a\}$. The computations were carried out using the software program Maps and Bifurcations, developed by Professor William A. Schaffer[15] of the University of Arizona. The ordinate is the normalized value of the variable, and hence $0 < x < 1$. The abscissa represents the variations in the map parameter $\{a\}$. Here the range $2.95 < a < 4.00$ was chosen. The program starts with $a = 2.950000$, where we know the fixed point to be stable.

In running the program the computations are performed for increasing values of $\{a\}$. In other words, as we move to the right, we move farther away from equilibrium. The numerical computations confirm one's intuition that as one moves away from equilibrium instabilities are likely to arise. We can see this in Fig. 10. Thus when $\{a\}$ reaches 3, a bifurcation occurs, i.e.,

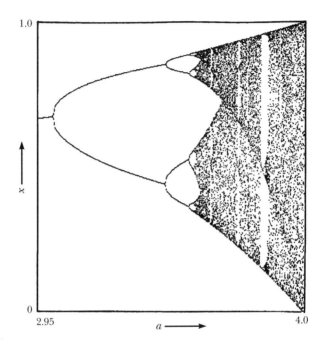

Figure 10. Iterates for the logistic equation for 2.95 < a < 4.0. (Computations performed with W. M. Schaffer's program.)

we now have two steady values. As {a} is increased, or as we move to the right, each of these branches bifurcates into 4, into 8, into 16, etc. One can note that the intervals between bifurcations decrease as {a} increases, culminating into a hash, i.e., chaos sets in. Within this regime are evident bands or windows. Inside these bands there are few points that can result in attractors (Rasband[16]). Also evident are dark streaks that arch through the chaotic regimes. The structure of the bifurcations diagram has been studied by Jensen and Myers,[17] who assert that the streaks are the mapping of the critical point $a/4$ through the various order iterates of the logistic equation.[18]

One of the characteristics of nonlinear dynamical systems is *self-similarity*. By this is meant that if a region of the map is magnified, copies are generated. We shall discuss this in greater detail in connection with scaling and fractals, where it is of particular importance. However, because self-similarity also occurs in bifurcation diagrams, I show in Fig. 11 the results of computations performed with the Schaffer software program already cited.

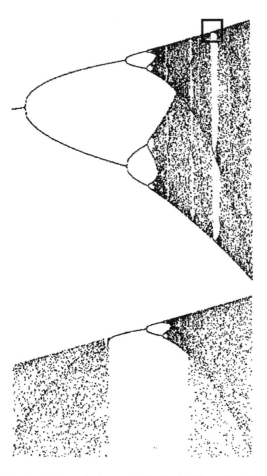

Figure 11. Magnification of edge of primary window in the bifurcation diagram. (Computed with W. M. Schaffer's program.)

The upper figure is a duplication of Fig. 10, except that a window box has been located at the upper edge of the band in the bifurcation diagram. The region contained within this window box is magnified with a convenient command in the Schaffer software, and the results are seen in the lower diagram of Fig. 11. The magnification shows a previously invisible bifurcation zone, and this is quite similar in form to the previous ones. The magnification also makes clear an earlier assertion, namely that the bands contain too few points for anything much to happen. The large window in the diagram occurs at approximately $a = 3.828427. . .$, and is believed to be stable.

In performing the computations yielding Fig. 10, one departs farther and farther from equilibrium as $\{a\}$ increases. As the evolution of the bifurcation diagram takes place, the phase diagram will change, and the successive time series will take on different forms. In Fig. 12 may be seen a montage of the various time series at different values of the map parameter $\{a\}$ super-imposed on the bifurcation diagram.

The excitement of observing on the computer monitor the unfolding of images of dynamical phenomena cannot possibly be duplicated with static text material, no matter how well illustrated. For those who cannot get access to the hardware and software I used in preparing the material covered in this chapter, but who nevertheless have access to a VCR player, I recommend a videotape developed by Professor R. L. Devaney[19] of Boston University.

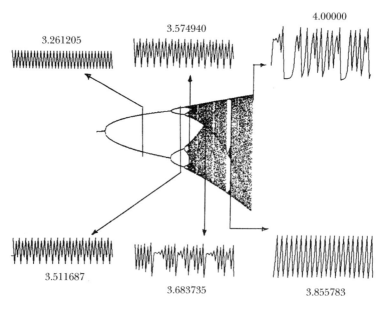

Figure 12. Montage of time series at different iterates of the logistic map. (Computations performed with W. M. Schaffer's program.)

FEIGENBAUM UNIVERSAL NUMBERS

As may be apparent by now from the bifurcation diagram, the values of $\{a\}$ at which bifurcations take place progressively come closer together. Please see Fig. 13. It would not be out of order to inquire if there is any pattern to this. The answer to this question was given by Rockefeller University physicist M. J. Feigenbaum[20], who showed that there is a universal number $\{\delta\}$, now named after him, that applies to different nonlinear dynamical equations. This is defined by the following relation:

$$\delta = \frac{a_n - a_{n-1}}{a_{n+1} - a_n} \tag{9}$$

The nomenclature is shown in Fig. 13.

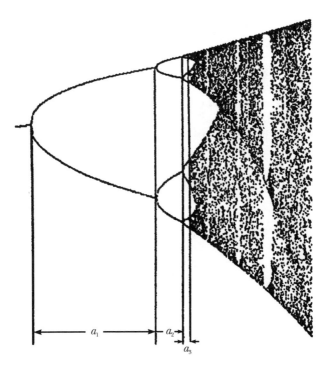

Figure 13. Selected locations of $\{a\}$. (Computation of bifurcation diagram performed with W. M. Schaffer's program.)

It is apparent that the different values of $\{a\}$ converge and that they do so geometrically. The values can be determined graphically, analytically, or computationally. Upon denoting the converged value by $\{a_\infty\}$ one can write together with Rasband[21] the following relation:

$$a_\infty - a_n = c / \delta^n \qquad (10)$$

where $\{c\}$ is a constant. Ultimately it is found that $c = 2.637\ldots$, $a_\infty = 3.5699456\ldots$, and the Feigenbaum number $\delta = 4.669202.\ldots$

Applications

The foregoing should serve to emphasize that the discrete logistic equation is a very powerful research tool. However, in applying it as a model to real-life problems, three restrictions must be realized. First, it applies to a single variable, and most complex dynamical problems occur exactly because a number of different factors interact. Second, there should be no overlapping of generations. This can be assumed such as in the case of insect species like the gypsy moth, where generations do not overlap. Third, reproduction does not require a pair of parents; only a single parent is necessary. In spite of these limitations, and although the derivations are based on population dynamics, the basic formulation serves as a useful model for a wide variety of applications, including genetics, epidemiology, and economics. For example, budgets and appropriations are made on a yearly basis, while in manufacturing a company might schedule its production by batch processes, or a medical laboratory might carry out certain diagnostic tests on certain days of the week. Of course, there are situations where generations do overlap. In economics these have been studied by Benhabib and Day.[22] When carefully applied, the Lotka–Volterra equation does indeed have diverse applicability. However, one must evaluate the circumstances before applying it. While it may not be elegant scholarship to fiddle around with negative results, I hope that the following example will drive home my words of caution.[23] (I feel comfortable because I am reminded of the statement made by the great Ludwig Boltzmann that matters of elegance should be left to the tailor and the cobbler.)

We asked the following question in 1988: "How well could one have predicted back in 1968 the future accumulation of solid waste in the U.S., using the discrete logistic curve?" At its face value the problem has considerable practical significance. From an environmental viewpoint one might want to know what waste accumulation is likely to be, and thereby initiate plans for land-use management, waste disposal, or alternative energy conversion. The computations of discrete logistic equation were performed by my former graduate student, W. F. Zeller III, using a spreadsheet approach, and are shown in Figs. 14a–14d.

The actual historical data were obtained from the 1987 U.S. Statistical Abstracts. A normalization of 1 = 8,000 was used because in the preliminary calculations it was found that in the oscillatory regime the maximum amount of solid waste would be 8,000 million tons. Thus for an initial actual value of 82.3 million tons, $x_0 = 82.3/8,000 = 0.0103$.

For purposes of practicality, the results were presented in the form of an actual time series and with actual waste quantities. As may be seen in Fig. 14a, the calculated values diminish with time for $a = 0.9500$, and this is completely in accord with the theory we discussed previously. However, as may be seen, the actual data show that just the contrary was occurring. The actual data and the calculated values are in reasonable agreement in Fig. 14b, where $a = 1.0397$. In Fig. 14c for $a = 3.600$, the theoretical calculations indicate the oscillatory process that is predicted, although the actual data hardly vary when plotted on this large scale. Finally, for $a = 4.05$, the calculated values indicate a negative accumulation of solid waste, which of

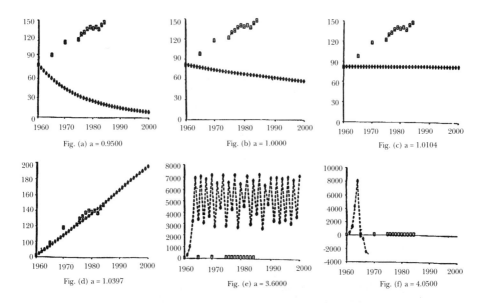

Figure 14. Solid waste accumulation. Diamonds are computed values, rectangles are actual data. (Computations performed by W. F. Zeller III, reproduced with his permission.)

course does not make any practical sense. Clearly, the discrete logistic curve chosen was not applicable to the problem at hand because solid waste is generated continuously, overlaps, and does not at all behave like the reproduction of insects! As they should have, the theoretically calculated curves behave exactly as one would expect them to do. All of this points out the importance of validating the calculations to whatever extent data may be available. It should be noted that the municipal waste problem has been studied more recently by S. S. Penner and M. B. Richards[24] who used the continuous form of the logistic equation and arrived at useful projections.

MULTIVARIABLE EQUATIONS

So far we have emphasized the behavior of a single variable, x. This should not be denigrated; the discrete logistic equation is generous with the general insights it provides. Of course, in many problems several variables interact. We discussed the treatment of many variables for the continuous case in the previous chapter. For the two-variable case, the discrete logistic equation is given by Jensen:[25]

$$x_{n+1} = x_n + y_{n+1} \tag{11}$$

$$y_{n+1} = y_n + k \sin x_n \tag{12}$$

where $\{k\}$ is indicative of the nonlinearity.

I should admit that the equations we discussed are far simpler than actual complex problems, with their demand for many more dimensions. Unfortunately, the state of the art of chaos theory has not advanced sufficiently to cope with many more dimensions. Still we should be happy because the field is expanding rapidly, and the modest beginning we have considered here is already providing insights that the traditional approaches did not reveal.

NOTES AND REFERENCES

1. Guckenheimer, J., and Holmes, P. (1983). *Nonlinear Oscillations, Dynamical Systems, and Bifurcations of Vector Fields*, New York: Springer-Verlag.
2. In complex dynamic phenomena one is frequently interested in projecting into the future. In doing so we rely on differential equations that express changes of one or more variables with time. If only one independent variable is involved, the differential equation is said to be an ordinary differential equation {ODE}, while when more than one independent variable is involved we must resort to partial differential equations {PDE}s. Differential equations can be classified otherwise. If the independent variable is known

at some time prior to the termination of the process, we deal with an initial value problem, {IVP}. On the other hand, if two different values of the independent variable are known, we have a boundary value problem, {BVP}. The logistic equation is an IVP.

3.   Li, T. Y., and Yorke, J. A. (1975). "Period Three Implies Chaos," *American Mathematical Monthly*, **82**, 985–992.

4.   Carnahan, B., Luther, H. A., and Wilkes, J. O. (1969). *Applied Numerical Methods*, New York: John Wiley & Sons.

5.   Jackson, E. A. (1989). *Perspectives of Nonlinear Dynamics*, Cambridge: Cambridge Univ. Press.

6.   Beltrami, E. (1987). *Mathematics for Dynamic Modeling*, Cambridge, MA: Academic Press.

7.   Penner, S. S., and Wiesenhahn, D. F. (1985). "Stability of Growth Rates in Energy Technologies," *Energy*, **10**(8), 983–985.

8.   May, R. M. (1976). "Simple Mathematical Models with Very Complicated Dynamics," *Nature*, **261**, June 10, 459–471.

9.   May, R. M., and Oster, G. F. (1976). "Bifurcations and Dynamic Complexity in Simple Ecological Model," *The American Naturalist*, **110**, 573–599.

10. May, R. M. (1976). *op. cit.*

11. Jensen, R. V. (1987). "Classical Chaos," *American Scientist*, **75**,(2) 168–181.

12. D'Amico, B. (1991). Reproduced here with permission, February 3, 1992.

13. Levitan, H. (1988). *Chaos and Fractals: Basic Concepts Demonstrated by the Logistic Model of Population Ecology*, College Park, MD: University of Maryland. Note: This program was written in the Windows environment. Site licenses may be obtained for classroom usage.

14. Holmes, P. (1984). "Bifurcation Sequences in Horseshoe Maps: Infinitely Many Routes to Chaos," *Physics Letters*, **104A**(6,7), 299–302.

15. Schaffer W. M. (1989). *Chaos in the Classroom: I, Maps and Bifurcation*, Leesburg, VA: Campus Technology.

16. Rasband, S. N. (1990). *Chaotic Dynamics of Nonlinear Systems*, New York: John Wiley & Sons.

17. Jensen, R. V., and Myers, C. R. (1985). "Images of Critical Points of Nonlinear Maps," *Physical Review A*, **32**, 1222–1224.

18. I am grateful to my former graduate student, S. R. Kasputis, for bringing this to my attention.

19. Devaney, R. L. (1990). *Transition to Chaos—The Orbit Diagram and the Mandelbrot Set*, New York: The Science Television Company.

20. Feigenbaum. M. J. (1979). "Quantitative Universality for a Class of Nonlinear Transformations," *Journal of Statistical Physics*, **19**, 25–52.

21. Rasband S. N. (1990). *op. cit.*

22. Benhabib, J., and Day, R. H. (1982). "A Characterization of Erratic Dynamics in the Overlapping Generations Model," *Journal of Economic Dynamics and Control*, **4**, 37–55.

23. Çambel, A.B. (1989). "An Overview of Self-Organization in Social Structures," *Dissipative Strukturen in Integrierten Systemen* (pp. 111–131), A.B. Çambel, B. Fritsch, and J.U. Keller, eds. Baden-Baden: Nomos Verlagsgesellschaft.

24. Penner, S. S., and Richards, M. B. (1989). "Estimates of Growth Rates for Municipal-Waste Incineration and the Environmental Control Costs for Coal Utilization in the U.S.," *Energy*, **14**(12), 961–963.

25. Jensen, R. V. (1987). *op. cit.*

CHAPTER **8**

THE DIFFERENT PERSONALITIES
OF ENTROPY

INTRODUCTION

Sir Winston Churchill's aphorism, ". . . a riddle wrapped in a mystery inside an enigma," applies well to the notorious *entropy function*. One cause for the confusion is that there are many versions of entropy, which is usually denoted by the letter {S}. For example, there is the original entropy due to Clausius, which is defined for isolated, macroscopic systems in equilibrium, undergoing reversible processes. In contrast there is the entropy of Prigogine that describes the behavior of open, dissipative structures. Boltzmann's statistical entropy and Shannon's information entropy are mirror images of one another, and this asymmetry causes grief to persons who seek absolute consistency. There is Eddington's "arrow of time," which expresses the direction of events and is a foundation stone for the indisputability of irreversible processes. And then there is Kolmogorov's entropy, which describes the deviations between the trajectories of dynamical systems and bears no resemblance whatsoever to Clausius's original definition.

Such a plethora of concepts bearing the same name is not unlike a family that through marriages has spawned a number of highly opinionated individuals who each have his or her own agendas. There is nothing wrong with the diversity, and it is encountered in science although we normally conceive of science as being very specific and categoric. For example, it was observed long ago that like Dr. Jekyll and Mr. Hyde, electrons have dual characteristics. In certain experiments they behave like particles, and in other experiments they move about in a wavelike manner. While in the early days of quantum mechanics this dualism was viewed as contradictory, the great Danish physicist Niels Bohr (1885–1962) pointed out that these two types of behavior were actually complementary. According to his *principle of complementarity,* physical phenomena may be described as waves or as particles, but not as both at the same time. As Werner Heisenberg[1] (1901–1976), a recipient of the Nobel Prize, pointed out:

> . . . the concept of complementarity introduced by Bohr . . . encouraged physicists to use an ambiguous rather than an unambiguous language, to use the classical concepts in a somewhat vague manner. . . . We realize that the situation of complementarity is not confined to the atomic world alone; we meet it when we reflect about a decision and the motives for our decision or when we have the choice between enjoying music and analyzing its structure.

Another great physicist, Louis de Broglie (1892–1987), who received the Nobel Prize for pointing out that particles can behave like waves,[2] came to the defense of ambiguity and equivocation:

> . . . we could hold, contrary to Descartes, that nothing is more misleading than a clear and distinct idea.[3]

By invoking the authorities of Bohr, de Broglie, and Heisenberg, I hope that blind adherence to absolute determinism and specificity are not practiced. Unless one tolerates some ambiguity it is difficult to gain insights into complexity and chaos. This leads to the entropy imperative.

Perhaps there is a little leeway in the case of entropy. One can evaluate, for the same system, different entropies that are not incompatible with one another, simply because they are not the same and differ in form, purpose, and numerical value (Grad[4]). The variety should not discourage us. As the 1977 Nobel laureate in chemistry, Ilya Prigogine,[5] expressed elegantly:

> Music is not exhausted by its successive stylizations from Bach to Schoenberg. Similarly, we cannot condense into a single description the

various aspects of our experience. We must call upon numerous descriptions, irreducible one to the other, but connected to each other. . . .

The connecting link among the different forms of the entropy is that, one way or another, they are indicative of deviation from equilibrium and chaotic behavior. There is no reason why they must look alike or act alike. I shall not attempt to artificially enforce consistency among the different definitions. I am quite content to embrace the principle of the complementarity principle. Those who insist on a philosophical justification are reminded of the American essayist Ralph Waldo Emerson (1803–1882), who stated, "A foolish consistency is the hobgoblin of narrow minds." I would like to be a little more scientific.

IS ENTROPY FOR REAL?

Even without being aware that there are quite a few types of entropy, the subject can give heartburn. One reason is that unlike properties such as temperature, we do not encounter it in our daily routines. Nor is there an entropy meter, in the convenient way that there are thermometers. Hence we do not internalize it the way we are accustomed to dealing with the more common metrics. This does not mean that the entropy cannot be measured. The third law of thermodynamics, sometimes recognized as Nernst's Heat Theorem,[6] makes this possible. It states that the entropy is zero at absolute zero temperature. The entropy is not so illusory after all.

We must realize that in actuality a number of common metrics—for example, temperature—are really mental constructs. This assertion deserves a little explanation. It is customary to define temperature as "the measure of hotness and coldness." To determine how much coldness or hotness exists, we use the concept of temperature, and this is manifested to us in the form of the length of the meniscus in a thermometer outfitted with a scale given in degrees Fahrenheit or Celsius (centigrade). We cannot place the concept called "temperature" into a box for safekeeping. While it is a property of systems, it is sort of ephemeral. This transitory nature of temperature does not bother us, because we are used to temperature changes whether it be by listening to a weather person on the news, or hearing the nurse murmur to the physician as we hopefully recover from illness. Clearly, temperature is not as tangible as is a piece of rock.

Let us be a little more curious and ask: When is a body really hotter than another? And the answer is: When its constituent particles are more highly agitated, namely when their kinetic energy is higher. But it is impossible to have particles that move about at a fantastic speed wait long enough to have their temperature taken. However, we can determine their energy by simple calculations[7] that tell us that about 10,000°K is equivalent to 1.29 electron volts. Once we accept this pluralistic way of looking at matters we realize that many concepts that we take for granted, and ascribe a physical meaning to, are really the fruits of our imagination that we have gotten used to. Bluntly stated we have been "programmed."

We could coin a definition for entropy: "Entropy is the ability to reach equilibrium." Another definition of entropy could be: "a measure of chaos." Still another could be: "an indication of transmitting information." This is why entropy, unlike energy, is not conserved.

There have been occasions where persons were willing to accept the concept of entropy and ask me which one they should use. My simple answer is: "There is no one best entropy definition. Use the one best suited for your purposes. Also, please don't overlook the possibility that describing certain complex systems may require evaluating several of its entropies, not just one."

The connecting link among the different forms of the entropy function is that, in one way or another, they are indicative of dissipative effects, deviation from equilibrium, information, and chaotic behavior. There is no reason why they must look alike as if they had come out of the same mold.

WHY MUDDY THE WATERS WITH ENTROPY?

One of my good friends, a colleague I esteem highly for his scholarship, raises his eyebrows whenever I bring entropy into our spirited discussions concerning chaos theory. I know he is not alone, so allow me to share with you, dear reader, the reasons why one should be mindful of entropy. First of all, modern deterministic chaos generally deals with dissipative structures, not conservative systems. These are not reversible, and the entropy serves as a compass to indicate the direction away from the equilibrium state. Second, complex dynamical systems involve uncertainty—namely, incomplete or missing information. Here Shannon's entropy proves to be a powerful measuring stick. Third, complexity involves randomness. Here Boltzmann's statistical entropy is indispensable because it deals with probabilities and is a measure of chaotic conditions. Fourth, complex dynamical systems follow trajectories;[8] their divergence is measured by the Kolmogorov entropy. Finally, complex systems are open and dissipative, and hence in addition to the internal entropy production, they experience entropy exchanges with other systems or their environment. This explains the operation of chemical and biological clocks, as well as other types of self-organization that are crucial to living beings. This is why Prigogine's entropy is so powerful.

It may be suggested that if one were to confine the discussion to conservative dynamical systems, one could get away without much attention to entropy, except perhaps an axiomatic mention here or there. However, this would be like knowing a person by name alone; it is not very satisfying. Furthermore, most of the problems of particular interest are dissipative. Because science advances inexorably, we learn more about the issues, and the more we learn, the more we inquire about the degree of complexity that exists. Those who must cope with complexity need to understand the rudiments of the different concepts that go by the same name, i.e., entropy. As Heisenberg put it:[9]

> . . . in the process of expansion of scientific knowledge the language also expands; new terms are introduced and the old ones are applied in a wider field or differently from ordinary language. Terms such as "energy," "electricity," "entropy" are obvious examples.

Entropy is inextricably linked to energy, information, and chaos. But what really are energy, information, and chaos? Let us concentrate on energy here, because the concept of entropy started with it; alternate forms of it are associated with information and chaos. Matter in the form of a natural rock or a manmade automobile is relatively easy to comprehend because it can be seen and felt. On the other hand, energy is ephemeral. We know about it indirectly because we see evidence of energy having been used or transformed. Energy is more of a concept than a thing. Yet we generally think that we know what energy is. There are several reasons for this state of mind: First, while carrying a heavy object, or engaging in sports, we can feel that we must exert an extra effort. We relate energy to personal qualities such as vigor. We envy persons who have strong physiques. We applaud energetic persons. We purchase gasoline to power our cars, natural gas to heat our homes, and electricity to operate our computers. We have no trouble accepting the concept of energy because we have become accustomed to it. It is not much different than the position of the fictitious professor, Henry Higgins, in the Lerner and Loew musical *My Fair Lady*. He admits to having gotten accustomed to the flower girl, Eliza Doolittle, while lamenting that he cannot understand her. It would behoove us to admit that in science, too, we cannot understand everything, and that uncertainties are unavoidable. We must abjure the arrogance of certainty.

Those who have taken a rudimentary course in science are indoctrinated with the textbook definition that "energy is the ability to do work." The brilliant Anglo–American physicist Freeman Dyson[10] compares the textbook definitions of energy with the poet William Blake's definition of energy in the latter's *The Marriage of Heaven and Hell,* published in 1793. Dyson writes:

> . . . when revolutions in thought have demolished old sciences and created new ones, the concept of energy has proved to be more valid and durable than the definitions in which it was embodied. In Newtonian mechanics energy was defined as a property of moving masses. In the 19th century energy became a unifying principle in the construction of three new sciences: thermodynamics, quantitative chemistry and electromagnetism. In the 20th century energy again appeared in fresh disguise playing basic and unexpected roles in the twin intellectual revolutions that led to relativity theory and quantum theory. . . . It is unlikely that the metamorphoses of the concept of energy, and its fertility in giving birth to new sciences, are yet at an end.

Originally, when first defined, entropy and energy were inextricably related. As Dyson points out, with advances in sciences energy has assumed different forms, and hence the character of entropy has changed, too. After the enunciation of Clausius, Boltzmann reformulated the definition of entropy from a statistical, probabilistic viewpoint. According to this, too, the entropy would be expected to increase. Enter the versatile British genius, James Clerk Maxwell (1831–1879). In 1871 he introduced the concept of a box having two compartments {A} and {B} with a two-way trap door between the compartments. This is the well-known "Maxwell's Demon," because a clever being would open the door in one direction or another so that the fast molecules (hot ones) would be shooed into one side of the partition, i.e., into part {B}, and the slow molecules (cold ones) into the other side of the partition, part {A}. In this way, Maxwell conjectured, the temperature of {B} would rise, while the temperature of {A} would fall without any expenditure of energy, and the second law of thermodynamics would be violated.

It was a clever ploy that many fell for, but Hungarian–American physicist for all seasons, Leo Szilard (1898–1964), pointed out that to perform his feat the Demon would have to be located inside the box and, furthermore, would have to act by expending information concerning which molecules would be hot and which ones would be cold. To be sure, the entropy would be reduced in the cold compartment, but this would be more than made up for elsewhere, particularly, as Brillouin explained, there would have to be a light source to enable the Demon to differentiate between the molecules. In the final analysis the second law would be violated after all, and irreversibility cannot be ignored. Both Szilard and Brillouin helped usher in the information interpretation of entropy that Claude Shannon perfected. Thus an informational interpretation of entropy was established. Now another era is awakening. This is the geometric interpretation of the entropy stimulated by the bizarre shapes that strange attractors have. We shall lay the groundwork for this in this chapter, but elaborate on fractal dimensions in a later chapter.

In describing the various persona of entropy, I shall classify entropy into the following groups: *macroscopic* (phenomenological), *statistical* (informational), and *dynamic* (geometric).

MACROSCOPIC ENTROPY

In this section I shall discuss the entropies named after Rudolf Clausius and Ilya Prigogine. The latter is an extension of the former, but it has revectored our thinking about dynamic systems. It was Rudolf Clausius who first conceived of the concept of entropy. Because of the severe restrictions imposed on it, the applicability of Clausius's entropy to complex systems is rather limited. However, I discuss it because of its "Godfather" status and the dominant influence it still maintains.

Clausius's Entropy

In 1850 Rudolf Clausius (1822–1888) presented his statement of the second law, which asserts that heat cannot rise from a colder to a hotter body without aid from the outside. In 1865 he arrived at the definition of the so-called entropy function, $\{S\}$.

$$dS = \left(\frac{\delta Q}{T}\right)_{rev} \tag{1}$$

In this definition, $\{Q\}$ is the thermal energy, and $\{T\}$ is the absolute temperature. Here I differentiate between the differential operators $\{d\}$ and $\{\delta\}$. The former denotes an exact differential, i.e., one whose integral is independent of the configuration of the path followed. This is because the entropy is a property. In contrast the latter, i.e., $\{\delta\}$, denotes an inexact differential because it depends on the type of process. This derives from the realization that heat or thermal energy, {Q}, is not a property. The term {1/T} serves as an integrating factor, which is the mathematical artifice of converting an inexact differential into an exact one. The notation {rev} at the bottom of the parentheses signifies that the definition of the entropy applies to reversible processes only. There are additional restrictions placed on this definition of the entropy. Specifically, the system must be isolated, it must be in equilibrium, and it must be macroscopic.

Cardwell[11] quotes Clausius:

> . . . I hold better to borrow terms for important magnitudes from the ancient languages so that they may be adopted unchanged in all modern languages, I propose the magnitude S, the entropy of the body from the Greek word τροπη, *transformation.*

How prescient these words have proven to be! As we shall see in the remainder of this chapter, the term "entropy" has endured numerous transformations.

Clausius further showed that entropy will increase in all irreversible processes that isolated systems undergo. This is expressed as follows:

$$dS \geq 0 \tag{2}$$

The equality applies to reversible processes, and the inequality to irreversible systems. Because reversible systems are not natural, the entropy tends to increase in general. Equation 2 states that the macroscopic entropy increases irreversibly, ultimately reaching its maximum. When this occurs the system will achieve its equilibrium state when its energy is minimal. Thus, it is evident that events proceed in the positive direction.

There has been considerable debate about the second law of thermodynamics and the applicability of the entropy function. For example, the entropy polemics[12] in 1903 among Heaviside, Lodge, Planck, Poincaré, and Swinburne were quite spirited. The debate was fueled significantly by two statements Clausius made categorically: 1) The energy of the universe is constant. 2) The entropy of the universe tends to a maximum. From the latter statement evolved the notion of the dreaded "heat death," namely, the eventual degradation of all useful energy of the universe. It was dramatic and caught the attention of people. Of course, this conclusion should not have been drawn, because it applies to isolated systems only, whereas the present scientific wisdom indicates that the universe is not isolated, but it is open and expanding (Weinberg[13]).

The general impression that entropy per se is deleterious is not entirely correct. This can be seen in the following example.

Figure 1. Rudolf Clausius (U.S. Library of Congress).

Example: As human beings we yearn for order, and this is natural because our lives would be so much more predictable. On the other hand, the orderly form of systems is not always the most useful or productive form. Consider a collection of H_2O (water) molecules. These can exist in three basic phases, i.e., solid (namely, ice), liquid (namely, water), and vapor (namely, steam). At a pressure of one atmosphere, water freezes at 0°C and evaporates at 100° C. When the water is frozen we have crystals of ice, which are arranged in an orderly fashion, whereas the water molecules making up the steam are in a state of constant agitation. In the liquid state the motion of the water molecules is some place between the orderliness of ice and disorderliness of steam. Summarizing, the ice crystals are quite orderly, the water molecules are less orderly, while the steam molecules are in a state of disorder or chaos.

As a matter of curiosity we inquire how the entropies of water and steam compare. These values of the entropy can be found in readily available tables. We find that at a temperature of 100°C, the entropy of water is 1.3062 kJ per kg per kilogram of water per degree Kelvin (kJ/kg-K), while again at 100°C, the entropy of steam is 7.3554 kJ/kg-K, obviously much higher. This confirms our physical intuition that steam is far more disorderly than water. However, this does not mean that steam is less useful. For example, we boil water for a variety of personal uses: making tea, doing the laundry, or operating the electric power plants that provide the electrical energy on which we so much depend. The moral is that we should not confuse a high entropy with undesirable circumstances.

As science advanced, the definition of Clausius's entropy proved to be confining for the study of complex dynamical systems, and major modifications were made.

Prigogine's Entropy

It has been pointed out by renowned gas dynamicist Ascher Shapiro[14] that classical thermodynamics deals with systems in equilibrium, and therefore the term "thermostatics" would be more descriptive because no dynamics is involved. This differentiation was amplified by Professor Myron Tribus.[15] Three stages of thermodynamics were described by Prigogine and Stengers[16] in a manner particularly applicable to complex systems, thermostatics, linear thermodynamics, and far-from-equilibrium thermodynamics.

Because we are concerned here with dynamical systems I shall not pursue thermostatics further. Hence we differentiate between two levels of non-equilibrium. One is *near to equilibrium,* and the other is *far from equilibrium.* The former is also called "irreversible" or "linear" thermodynamics. It was developed by Lars Onsager (1903–1976), the recipient of the 1968 Nobel Prize in chemistry. It applies to transport phenomena such as diffusion, and electrical and thermal conduction. These flows (fluxes) obey linear relations because the flows are linear functions of the forces (potentials). The fundamentals of linear thermodynamics are embodied in the *reciprocity relations* of Onsager.[17]

In a wide variety of complex dynamic systems, the flows are nonlinear functions of the forces. Under these circumstances we speak of "far-from-equilibrium thermodynamics." It is far-from-equilibrium thermodynamics that is useful in the study of self-organizing systems. Shock waves are another example. The pioneering work here has been formulated by Prigogine[18,19] and his associates (Glansdorff and Prigogine,[20] Nicolis and Prigogine[21]).

Figure 2. Ilya Prigogine (News and Information Service, the University of Texas at Austin).

What Prigogine did was to extend Clausius's entropy to open systems. As we well know in real-life situations, there are inflows and outflows such as matter, energy, information, and entropy between open systems and their environment. Systems that experience an increase of entropy are called *dissipative structures*. Prigogine showed that dissipative structures can lead to self-organization. This implies that more complex structures can evolve from simpler ones. One can therefore conceive of order coming out of chaos. Indeed, this is the title of the best selling volume by Prigogine and Stengers.[16]

Clausius had presented the relation that accounts for the entropy increase in an isolated system. Prigogine elaborated on this so that not only the entropy increase within the system as demanded by Clausius, but also the entropy exchange between the system and the surroundings, could be determined. His celebrated relation, which applies to all systems, is written:[21]

$$dS_T = dS_I + dS_E \qquad (3)$$

Here $\{dS_T\}$ is the net or total entropy change, $\{dS_I\}$ denotes the component of the entropy change inside the system, namely the entropy after Clausius. This is the entropy that tends to increase, and cannot decrease. The term $\{dS_E\}$ represents the part of the entropy that the system exchanges with its surroundings, and this can occur only in open systems. Whereas $dS_I \geq 0$, $\{dS_E\}$ may be positive, negative, or zero. It would be positive if entropy enters the system, and negative if entropy is discharged to the outside like putting out the garbage. Of course, $dS_E = 0$, if the system were reversible as well as isolated. Therefore, $\{dS_T\}$ can be positive, negative, or zero. The sign depends on the relative magnitude of the dissipative entropy change within the system, and the magnitude and the algebraic sign of the entropy exchange.

If the absolute values of the two entropies are such that $|dS_E| > |dS_I|$ but the system rejects entropy so that $dS_E < 0$, then the net entropy change $dS_T < 0$. This means that we remove a sufficient amount of entropy, or stated differently, we provide a sufficient amount of negative entropy, the latter being a manifestation of order. In other words, the system discards chaos, noise, or whatever we feel uncomfortable with. One way of looking at it is discarding useless items while cleaning house. Another metaphor is aid to needy areas (Cambel[23]). Consider a poor area that is going from bad to worse because of poor nourishment, ill health, corruption, or job loss. Clearly the internal entropy will increase, i.e., $dS_I > 0$. An infusion of aid in the form of expertise, necessary basic goods, and funds could conceivably change the situation, because for such items $\{dS_E\}$ would be negative. Obviously, $\{dS_E\}$ would have to be provided at the proper rate as persons familiar with foreign aid and economic redevelopment well know.

Previously, we noted that the entropy of isolated systems tends to increase; at best it can remain the same if the isolated system undergoes a reversible process. Structures for which $dS_T < 0$ are said to be *self-organizing*.[25] Rewriting, we have $dS_I + dS_E = dS_T < 0$, or $dS_E = -dS_I < 0$. This must occur under conditions of nonequilibrium, because otherwise the two entropy change components would vanish, and this of course would be a trivial case. It follows that nonequilibrium can result in order. But not all nonequilibrium situations result in order. On the contrary, some get out of hand and become chaotic. It depends on what type of feedback mechanism is operational. Thus the self-organizing capability of structures may be modified. For example, physiologist Friedrich Cramer,[26] of the Max Planck Institute for Experimental Medicine, has suggested that a living system is a highly complex limit cycle, responsive and adaptable to environmental changes that are self-organizing, but that the vicissitudes of aging are due to loss of symbiotic capability, loss of regulatory capability, and change in proteins.

Entropy Balance Equation

Whenever there is a flow, physical, chemical, or socioeconomic changes occur. These are described by so-called equations of change, or balance equations. Such equations differ from conservation equations in that some kind of production, or annihilation, occurs. While energy is conserved, entropy is not. To determine the entropy flow into or out of a system, an entropy balance equation must be written. The system exchanging entropy with its surroundings is represented by a volume element $\{v\}$. A balance equation for entropy may then be formulated (Yourgrau, van der Merwe, and Raw[27]) as follows:

$$\left|\begin{array}{l}\text{Rate of entropy change}\\\text{within system volume}\end{array}\right| + \left|\begin{array}{l}\text{Net entropy flux}\\\text{across boundary}\end{array}\right| = \left|\begin{array}{l}\text{Entropy production}\\\text{within system volume}\end{array}\right|$$

The entropy balance equation follows from the previous equation:

$$\frac{\partial(\rho s)}{\partial t} + \nabla \cdot J_S = \dot{\sigma}_s \tag{4}$$

Here $\{\rho\}$ is the density, $\{s\}$ is the entropy per unit volume, and $\{J_s\}$ is the entropy flux density, namely the entropy crossing per unit area per unit time. It is a vector quantity and hence its dot product with the vector operator nabla $\{\nabla\}$ yields a scalar quantity. Finally, $\{\sigma_s\}$ is the rate of entropy production. For very regular flows, such as a laminar flow, one would not expect any significant entropy production, whereas for turbulent flows there will be entropy production.

Gibbs Free Energy and the Quality of Energy

It follows from the second law that some of the energy supplied is not available for doing work. Anyone who has touched an automobile tailpipe can attest to its high temperature. The chemical energy of the fuel is converted into thermal energy. Some of this thermal energy is converted into mechanical energy, which appears at the wheels for locomotion, but some of the thermal energy is discarded into the atmosphere. This is not unlike a persons's paycheck. Not all of the money earned is available to spend as we please. Some must be paid to cover taxes. Therefore, one may write the following relation:

Available energy = Total energy supplied − Unavailable energy.

In the arena of communications one may write an analogous relation:

Information = Signal − Noise

The great American scientist Josiah Willard Gibbs (1839–1903) showed that the "free energy" during a chemical process is given by the celebrated Gibbs function, which is defined as follows:

$$G = I - TS$$

Here $\{G\}$ is the Gibbs free energy, $\{I\}$ is the enthalpy and represents the energy, $\{T\}$ is the absolute temperature, and $\{S\}$ is the entropy. It is evident from this equation that there is a sort of tug of war between the energy and the entropy. Entropy increases are accompanied by decreases of the Gibbs free energy. Thus there is a definite relationship between energy and entropy.

The first law insists that the "quantity of energy" be maintained, but is oblivious to the "quality of energy." Further consideration will lead us to the observation that some types of energy have a higher value or quality. For example, electricity can do more than heat water. Freeman Dyson[28] has arranged different forms of astronomical energy sources according to an "order of merit" that follows from their respective entropy to energy ratios. The lower this ratio, the higher the quality of the energy source. According to Dyson, the entropy per unit energy ratio is 0 for gravitation, 10^{-6} for nuclear reactions, and is in the range 1–10 for chemical reactions. This explains why hydroelectric energy is so efficient, and why combustion devices such as furnaces or automobile engines have relatively lower efficiencies.

STATISTICAL ENTROPY

So far we have considered the entropy from a phenomenological viewpoint. Now we shall look at it from a statistical viewpoint. The foundations of classical thermodynamics and statistical thermodynamics are very different. Classical thermodynamics is phenomenological, whereas statistical thermodynamics is a formalism that derives from mechanics. The entropy constitutes the bridge between them. This assertion is due to the observation that the statistical entropy assumes differential forms having the same basic structure as in classical thermodynamics. As Gibbs put it so succinctly:

> The laws of thermodynamics may easily be obtained from the principles of statistical mechanics, of which they are the incomplete expression.

Boltzmann's Entropy

Statistical entropy was originated by the great Austrian physicist Ludwig Boltzmann (1822–1888). He was determined to reconcile the reversible characteristic of classical Newtonian mechanics, which is time-invariant, with the formulations by Rudolf Clausius and William Thompson (Lord Kelvin) of the second law of thermodynamics, which insists on directionality and is time-dependent. In his efforts to do so, he was vociferously attacked by such formidable scientists as Pierre Duhem, Josef Loschmidt, Ernst Mach, Wilhelm Ostwald, Henri Poincaré, and Ernst Zermelo. They rejected his microscopic (atomistic) views because they held that physical theories should be formulated from macroscopically measurable quantities. Depressed and in poor health, he committed suicide. However, he was vindicated later by the revelations regarding Brownian motion and the acceptance of the atomic nature of matter. There is no doubt that we owe it to Boltzmann that we do not blindly accept the time invariance inherent in the laws of physics.

While irreversibility was acknowledged on the macroscopic level (who could deny aging?), it was Boltzmann who introduced the concept of time direction on the microscopic level. It was his insistence on asymmetric time that aroused his attackers. This contention was a direct result of the celebrated Boltzmann equation, which has stimulated many great scientists to make refinements on it. For the relatively simple case of transport in gases, the Boltzmann equation may be written as follows:

$$\frac{\partial f}{\partial t} + v \cdot \frac{\partial f}{\partial \vec{r}} + F \cdot \frac{\partial f}{\partial \vec{v}} = \left(\frac{\partial f}{\partial t}\right)_{coll} \tag{5}$$

Here f represents the distribution function $f(\mathbf{v}, \mathbf{r}, t)$, $\mathbf{v}$ and $\mathbf{r}$ are the velocity and position vectors, respectively, and $\mathbf{F}$ is the force, which is a function of $\mathbf{r}$ and t only. The terms on the left side of Eq. 6 account for the motion of the particles, while the term on the right side is the "collision term" dealing with the collisions among particles. The Boltzmann equation is asymmetric in time.

Boltzmann was then led to his famous H-theorem, which describes the tendency of particles to move towards their equilibrium distribution regardless of their initial distribution. In other words, the quantity $\{H\}$ is a measure of how much a system deviates from its equilibrium condition. This quantity is defined

$$H = \sum p_i \log_e p_i \tag{6}$$

Boltzmann further found that this quantity must decrease over time as matters evolve. Thus

$$\frac{dH}{dt} \le 0 \tag{7}$$

whereas we had seen in Eq. 3 that the quantity $\{S\}$, called the entropy, must increase. Otherwise there is quite a resemblance.

It was actually Max Planck who proposed the expression for the Boltzmann entropy, which is engraved on his tombstone in the central cemetery in Vienna. The expression is

$$S = k \log W \tag{8}$$

where $\{W\}$ denotes the microstates accessible to the system. The more states available, the higher will be the entropy. By now we can probably smell that there is some relation between $\{S\}$ and $\{H\}$, and indeed there is. Specifically, $S = -kH$ where $\{k\}$ is the Boltzmann constant. It follows that the correct expression of the statistical entropy is:

$$S = -k \sum p_i \log_e p_i \tag{9}$$

where p_i is the probability. The greater p_i, the higher the entropy. In other words, the probability of accessible states is an indication of uncertainty. There is opportunity for confusion because it depends on whether or not quantum mechanics was considered in the derivation. It is helpful to be aware of the H-theorem.

Let us make a mental leap. We know that particles are in random motion and that their behavior can be represented by a probability distribution.

Figure 3. Ludwig Boltzmann (Austrian Cultural Institute, New York).

This suggests that probability is a measure of randomness. Because this implies that entropy is probabilistic, it can be considered a measure of randomness or chaos. Looked at it in this way the entropy should be no more intimidating than the temperature.

Compared to the Clausius entropy, the statistical entropy is less restrictive. While Boltzmann started off with an isolated system to get at the H-theorem, the final expression for the entropy is quite general because it is not based on any particular particle kinetics, i.e., it does not require microscopic equilibrium. This is why the statistical entropy can be used as a measure of disorder, randomness, and chaos.

If there is more than one system, the entropies are additive:

$$S = S_1 + S_2 \tag{10}$$

and hence because probabilities are multiplicative,

$$W = W_1 W_2 \tag{11}$$

One of the most important aspects of the Boltzmann entropy is that it is an indication of disorder, and hence a measure of chaos. In effect this means that it is an indication of lack of information, which leads us to Shannon's information theory, hence uncertainty. The greater the uncertainty, the higher will be the number of complexions (structures), which leads to higher probability, and hence to a higher statistical entropy. Of course, uncertainty is one of the major aspects of complexity.

Brillouin-Schrödinger Negentropy

Another reinterpretation of the entropy came from Erwin Schrödinger,[29] the 1933 Nobel laureate in physics. He argued that living organisms feed on negative entropy, or *negentropy*.

> How would we express in terms of statistical theory the marvelous faculty of a living organism, by which it delays the decay into thermodynamic equilibrium (death)? It feeds upon negative entropy, attracting as it were, a stream of negative entropy upon itself, to compensate the entropy increases it produces by living and thus to maintain itself on a stationary and fairly low entropy level.

Schrödinger rewrote the statistical entropy equation in the following form:

$$S = k \log D \tag{12}$$

wherein he called $\{D\}$ the *disorder*. Brillouin[30] pursued this equation in greater detail, which was not unreasonable, because $\{W\}$ is the number of accessible states and, as we noted, is a measure of uncertainty. Schrödinger argued if $\{D\}$ is disorder, its reciprocal $\{1/D\}$ represents *order*. He then proceeded to write

$$-S = k \ \log(1/D) \tag{13}$$

But the logarithm of $\{1/D\}$ is the negative of the logarithm of $\{D\}$. Hence $-(\text{entropy}) = k \log(1/D)$. This negative entropy is the so-called negentropy, and it is meant to be a measure of order.

In essence Schrödinger was asking the right question regarding the entropy changes in living organisms, and his explanation to some degree is insightful. However, what was lacking is any emphasis that living beings are open systems.

After World War II the interest in entropy took on an entirely new direction. Entropy became related to information. Among the pioneers on this path was the brilliant American mathematician Norbert Wiener (1894–1964), who had been working on the theory of messages. He realized that his work had wider applicability, such as the control of machinery and society, computers, and the nervous system. He called this field "cybernetics."[31] In Wiener's words:[32]

> This larger theory of messages is a probabilistic theory, an intrinsic part of the movement that owes its origin to Willard Gibbs. . . .

Boltzmann had built the bridge between mechanics and thermodynamics, and Wiener outlined the bridge between probabilistic physics and information theory. Here was the connection to the statistical entropy that Boltzmann, Gibbs, Maxwell, and Planck had made respectable. In this era where allegations about lack of scientific integrity are so common, Wiener and Shannon are proof of the contrary. In his writing, Norbert Wiener went out of his way to acknowledge the benefit he had gained from Shannon's work, while Claude Shannon made special mention of the benefits he had derived from Wiener's contributions. Scientific sportsmanship at its best!

Shannon's Entropy

In thermodynamics entropy is not unlike taxes in everyday life; it tends to go up. And just as taxes reduce disposable income, entropy reduces available energy. In signal processing, such as in telephone communications or radio and television broadcasting, entropy is quite similar to noise. It reduces or compromises the signal. However, there are some fundamental differences between thermodynamic entropy and information entropy.

The impetus for the Clausius entropy originated with Sadi Carnot wanting to know the maximum efficiency that a steam engine could have. The impetus for what we now call Shannon's entropy is due to the Bell telephone system having to know how to cope with the ever-increasing number of telephones in homes and institutions. In 1948 Shannon published two seminal papers[33,34] in which he detailed the mathematical theory of communications, namely how messages are transmitted, the optimum manner of doing so, as well as how the signal bearing the message may be degraded.

It must be realized at once that "information" does not imply "meaning." As Shannon himself cautioned (Ref. 26, p. 31):

> [The] semantic aspects of communications are irrelevant to the engineering problems. The significant aspect is that the actual message is one *selected from a set* of possible messages. The system must be designed to operate for each possible selection, not just the one which will actually be chosen since this is unknown at the time of design.

Several observations follow from these remarks. It is possible that two messages may be technically identical, yet one may have meaning while the other may be nonsense. Second, there is no a priori assurance as to which of a number of messages sent will be received; in other words, choice is important, all choices being equally probable. From this follows the probabilistic or stochastic nature of the information. This led Shannon to follow Hartley's suggestion that in such cases the logarithmic function is useful. However, Shannon used logarithms to the base 2 because in modern communications one deals with binary digits, namely bits. Thus the two stable conditions are 0 and 1. A device then stores one bit of information at any one time. Because N devices can store N bits, the total number of states is 2^N.

$$\log_2 2^N = N \tag{14}$$

It follows that the logarithm of the number of signals increases linearly with time. Shannon then considered a set of possible events whose probabilities of occurrence would be $p_1, p_2, \ldots, p_n$, but one cannot know which event will take place. He inquired if there would be a measure $\{H_S\}$ that would give the uncertainty of the outcome. He then specified the requirements that such a measure should have, and showed that these would be possible only if $\{H_S\}$ is written as follows:

Figure 4. Claude Shannon (The M.I.T. Museum).

$$H_S = -K \sum_{i=1}^{n} p_i \log_2 p_i$$

(15)

This is the so-called Shannon uncertainty, Shannon's entropy, or the information entropy. Its resemblance to the statistical entropy is striking.

The constant $\{K\}$ does not automatically mean the Boltzmann constant $\{k\}$. Also, the logarithms have different bases.[35] It can then be shown (Andrews[36]) that

$$S = \frac{k}{K} H_S \qquad (16)$$

In the Boltzmann version of the H-theorem the probabilities are based on statistical averages because with so many particles it is impossible to know the behavior of individual particles. In contrast, in the Shannon version the probabilities indicate the choice that one message rather than another will be chosen. This can give us clues about the outcome of complex systems that by definition involve uncertainty.

Shannon elaborated on his equation further and presented a number of conditions. Among these the following are of particular importance when dealing with uncertainty. First, if we are certain of the outcome, $H_S = 0$. This can happen only if all p_i's are zero, except one. In contrast the most uncertain situation occurs when all of the p_i's are equal, because $\{H_S\}$ is then a maximum. If there are two linked events, the uncertainty of the joint events is less than or equal to the sum of the individual uncertainties. The information transfer will be maximum when the receiver "learns" something new. Information should be new in order to be informative.

In choosing to send one message or another, the quantity of information plays a key role. The greater the information, the greater the choice to select from among different messages that can be sent, and hence information is associated with freedom of choice. But if there is freedom of choice, uncertainty exists. We also know that entropy is a measure of information because it is a measure of randomness. But randomness, too, implies alternatives, choice, and uncertainty. Hence information and entropy are related. Unfortunately, this is sometimes misunderstood because it is felt that if the entropy is high, information must be low. In actuality the higher the probability, the greater the choice, and hence the chances of getting information through one way or another are enhanced. Entropy must be exonerated from the notoriety that over the decades has been ascribed to it. It had a bum rap.

The aforementioned overview of Shannon's equation focuses on discrete signals in the absence of noise. In real-life situations, continuous messages are very common, and as a rule noise cannot be avoided. Fortunately, it turns out that these factors do not conceptually alter the aforementioned basic approach, which can be modified to suit different circumstances. In his original papers Shannon described how one deals with noise and handles continuous messages.

Having noted that the log term in the statistical entropy is to the base $\{e\}$, it is useful to be able to change from one base to another. Suppose that we want to convert the logarithm of a number $\{N\}$ from base $\{a\}$ to base $\{b\}$. The transformation is as follows:

$$\log_b N = (\log_a N) \times (\log_b a) \tag{17}$$

Example What is the conversion of the decimal entropy to the binary entropy?

Solution $\log_{10} S = (\log_2 S) \times (\log_{10} 2)$, and

$\log_2 S = 3.3 \log_{10} S$

It follows that we can take conceptual advantage of the mathematical similarity between the Boltzmann and Shannon entropies, but we must be careful in numerical calculations.

There is a classic story of how Shannon came to name his equation the "entropy," instead of the "uncertainty." According to an anecdote of Shannon's, retold by Tribus:[37]

> When I discussed it with John von Neumann, he had a better idea. von Neumann told me: You should call it entropy, for two reasons. In the first place your uncertainty function has been used in statistical mechanics under that name, so it already has a name. In the second place, and more important, no one knows what entropy really is, so in a debate you will always have the advantage.

As Shannon pointed out, he focused on the engineering problems, not the semantic ones. While this is useful, I would be callous if I left the impression that semantics, or the meaning of the information process, are irrelevant. Far from it. Many of our problems are doubly complex because the meaning of the message is not well understood. Obviously information sent to a robot is far more primitive than the message that we want to convey to a professional associate. In the latter the structure of the language being used provides the means of injecting meaning to information. An apocryphal story illustrates this.[38] It goes something like this: In order to prevent misunderstandings due to translation from one language to another during important international negotiations, it is decided by United Nations officials to commission a translating computer. The computer is built and presented to an august group of linguists. As a test they suggest that it translate the English expression "out of sight, out of mind," into language X and then back into English. When the command for the reverse translation is given, these words appear on the screen: "Invisible idiot!"

Ergodicity

One of the many concepts that hark back to the early days of statistical mechanics is *ergodicity*. Both Boltzmann and Gibbs were eager to figure out how statistical ensembles might evolve over time from nonequilibrium to equilibrium in accordance with the laws of probability. They used Gibbs' ensemble idea because macroscopic continuum systems are made up of numerous particles, typically of the order of 10^{23} or more. When dealing with such large numbers of particles one cannot, of course, concentrate on individual particles. To overcome this complication Gibbs suggested that one consider a number of small systems that are configured identically. These are called an *ensemble*. The number of particles in an ensemble is finite.

Boltzmann formulated the ergodic hypothesis in 1871. Since then the study of ergodicity has followed two routes. One is the physical problem, the other is a branch of mathematics and measure theory. With the accelerating interest in complexity and chaos, the two viewpoints are starting to overlap. The details are beyond the scope of this introductory volume. However, I cannot use this observation as an excuse and ignore the subject. I shall mention it only briefly. Those who desire erudite treatments might consult the contributions of Eckmann and Ruelle,[39] Lebowitz and Penrose,[40] Ornstein,[41] and Petersen.[42]

There are several reasons why ergodicity is mentioned here. It is related to statistical entropy and information entropy, which we have just discussed. In later chapters we shall be concerned with the measurement and dimensioning of strange attractors that are associated with chaos and, in turn, with complexity. Another reason for being cognizant of ergodicity is its bearing on Arnold's cat, or the baker's transformation, and the Bernoulli shift. It is also relevant to Lyapunov exponential functions and to the Kolmogorov entropy, which we shall discuss in the next section. And, of course, ergodicity is significant whenever one deals with dissipative structures.

Ergodicity is a property of a system that provides a means to study the statistical behavior of dynamic systems as the system evolves according to probability theory. It is indicative of irreversibility. For example, the simple harmonic oscillator is ergodic. Unlike a deterministic system, an ergodic system is independent of its initial position. The ergodic principle can help us with the long-term, nontransient behavior of systems. Complexity occurs under conditions of nonequilibrium. It is natural to ask how the system may reach equilibrium.

DYNAMIC ENTROPIES

Throughout our discussion so far the underlying thread is that the entropy is indicative of irreversibility and randomness. In the following section I shall try to reveal the dynamic characteristics and the time dependence more explicitly.

Our preoccupation with time is an old one. Aristotle (B.C. 384–322) defined[43] it as, "Time is the measure of change with respect to before and after." This is a succinct definition of time, but there are a number of nuances that have continued to puzzle philosophers and scientists alike. Some of these occur because there are different types of time, and time can be measured in different ways depending on the observer and the observed. Certainly, the time scale of a geologist studying long-term geological changes taking place over millions of years is quite different from that of a high-speed photographer having to stop events that take place in nano-seconds (billionths of a second). Children and adults have different perceptions of time. Henri Bergson[44] (1859–1941) differentiated between the time that is a coordinate in science and the time we experience consciously. Implicit to this psychological time is that while we can remember the past we cannot recall the future. This, Bergson argued, constitutes an irreversible, one-dimensional flow that is dynamic. The dynamic nature of time is most fundamental to any evolutionary process. Yet our acceptance of it is ambivalent. Aristotle himself realized that motion may cease, but time does not.

According to Brillouin,[45] Lord Kelvin referred to time and irreversibility:

> If, then, the motion of every particle of matter in the universe were precisely reversed at any instant, the course of nature would be simply reversed forever after. . . . And if, also, the materialistic hypothesis of life were true, living creatures would grow backward, with conscious knowledge of the future but with no memory of the past, and would become again, unborn.

Eddington and Time's Arrow

It was Sir Arthur Eddington[46] who in 1928 coined the term "time's arrow," relating time to the second law of thermodynamics. His observation was that because entropy increases with time, it provides a means to differentiate between the past and the future.

We recall the past, but we can only guess about the future. The meteorologist Lorenz found out scientifically that long-term predictions cannot be made.

Figure 5. Arthur Eddington (U.S. Library of Congress).

It is the second law of thermodynamics that tells us anything about the direction of time, and this accords it a unique place among the laws of physics. Twelfth-century Persian poet and algebraist Omar Khayyam[47] wrote:

> The Moving Finger writes; and, having writ,
> Moves on: not all thy Piety nor Wit

Shall lure it back to cancel half a Line,
Nor all thy Tears wash out a Word of it.

Referring to the classical and modern laws of physics, Eddington remarked:

The classical physicist has been using without misgiving a system of laws
which do not recognize a directed time; there is only one law of nature—the
second law of thermodynamics—which recognizes a distinction between past
and future . . . [it] holds, I think, the supreme position among the laws of
Nature. . . . Let us draw an arrow arbitrarily. If as we follow the arrow we find
more and more of the random element in the state of the world. . . .

While Eddington pioneered in relating time, irreversibility, and randomness, his enunciations would require some fine-tuning in light of the more
recent developments.

It has been pointed by Hawking[48] that there are at least three arrows of
time: the thermodynamic arrow, namely, Eddington's; the psychological,
namely, Bergson's; and the cosmological arrow of time, which is the direction of time of the expanding universe. Hawking has further explained why
the three arrows must point in the same direction.

To these we should add a socioeconomic time, what Murphy[49] calls the
entropy time, because there is a difference between chronological time and
human time. As an example, consider the three major religious Sabbath
days, Friday, Saturday, and Sunday. Thus, if one is engaged in international
trade there are really four common working days in a week. Time zones,
too, cause socioeconomic disparities. Last but not least, time appears
differently to persons of different age groups and to persons in different
occupations. Plants and animals have different time horizons. For example,
a human's life span is measured in decades, but the sequoia tree may exceed
100 or 200 years. The gestation period of mammals differs, as well as the
age when they start reproduction. Obviously, time and the level of complexity are related, but we do not know how.

Much thinking has gone into thinking about time, but a good understanding remains highly elusive. An authority on time, J. T. Fraser, named
one of his books *Time, the Familiar Stranger.*[50] Maybe for the time being
(no pun intended), graffiti found in Austin, Texas, (Boslough[51]) sums it
up well, "Time is nature's way of keeping everything from happening all
at once."

At the time I wrote this section, a stimulating volume by Coveney and
Highfield[52] dedicated to the subject of the arrow of time appeared. It goes
far beyond my essay. I heartily recommend it to the peripatetic reader.

The Kolmogorov Entropy

If two populations, otherwise identical, vary by a minuscule difference at the onset of the time series, they will appear to follow the same trajectory at the beginning but then the deviation will become noticeable, and their respective trajectories will diverge. We saw confirmation of this when a slight change in the initial condition of the discrete logistic equation led to different trajectories. This implies that long-term predictions cannot be made with confidence, because information is lost as the entropy increases. To retrieve the original precision, more information must be provided. This is the informational counterpart of dissipative systems that we saw earlier, namely those systems that in order to continue functioning must receive energy from the outside. The Kolmogorov entropy is the measure of the information loss in N-dimensional phase space. We divide the phase space into cells, or hyperspaces, of size $\{\varepsilon\}$, and take measurements at time increments, $\{\tau\}$.

The trajectories move from cell to cell, and in order to predict in which cell they will be next, we need additional information. This is given by the difference of the Kolmogorov entropies, $K_{n+1} - K_n$, which in turn are expressed by the statistical entropies. The average rate of information loss is the Kolmogorov entropy defined in the following manner:

$$K = -\lim_{\tau \to 0} \lim_{\varepsilon \to 0} \lim_{N \to \infty} \frac{1}{N\tau} \sum_{i=0}^{N-1} p_i \log p_i \tag{18}$$

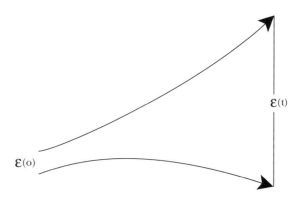

Figure 6. Exponential separation of trajectories in phase space.

Several possibilities exist. First, if the trajectories do not diverge, then $K = 0$, and there is no loss (or gain) of information. Second, if the trajectories of the initially adjacent points are exponentially separated, i.e., $N = e^{\lambda}$, where $\{\lambda\}$ is the Lyapunov exponent, then we have chaotic conditions (Schuster[53]). For this situation, $K > 0$. Finally, if the points are distributed with equal probability then we have random motion, and the Kolmogorov entropy is infinite (Baker and Gollub[54]). It follows that the Kolmogorov entropy can be useful in determining the onset of chaos.

Figure 7. Andrei N. Kolmogorov. (The Bettmann Archive).

We saw that the Boltzmann entropy serves as the bridge between mechanics and thermodynamics. Norbert Wiener designed the bridge between probabilistic physics and information theory. In turn the Kolmogorov–Sinai entropy serves as a bridge between information theory and geometry. Interpreted in this spirit of diversity, entropy constitutes a measure of chaos and should be considered as a dimension rather than a physical object.

Rényi Entropy

At this point let us step back for a moment and survey how chaos and complexity might be measured. We presented some approaches as early as Chapter 2, but promised to return for additional details. We did so in this chapter wherein we concentrate on entropy and notice that there has been an ebb and flow as it changed character, but that it remained loyal to its heritage as a measure. The discussion does not end here, however. In the next chapter when we discuss scaling and self-similarity in connection with fractals, we shall learn that there are still other dimensions, commonly called "fractal dimensions," the most well known being the Hausdorff–Besicovitch dimension. It follows that with all of this variety, a generalized dimension should be useful. This is what Rényi[55,56] attempted to do. He generalized the probabilistic entropy (please recall the spirit of Eq. 9 in this chapter) by writing the following expression:

$$S_q = \frac{1}{q-1} \, \log \sum_{i=1}^{N} p_i^q \tag{19}$$

for moments of order $\{q\}$ for the probabilities $\{p_i\}$ when $q \neq 1$ and is not necessarily an integer. Because by definition fractals have noninteger dimensions, this equation is applicable to fractals. When $q = 1$, this equation reduces to the well-known statistical entropy equation, namely,

$$S = -\sum_{1=1}^{N} p_i \, \log p_i \tag{20}$$

It can be shown (Farmer,[57] Grassberger,[58] and Schroeder[59]) that Eq. 19 is an information dimension and in harmony with the definition of the Hausdorf–Besicovitch fractal dimension with which we shall become familiar. The bridge to fractal dimensions is ready. Where does all of this leave Rudolf Clausius, who started it all? His place in history is secure. Please remember that he had chosen the word "entropy" which in Greek means "transformation."

NOTES AND REFERENCES

1. Heisenberg, W. (1958). *"Physics and Philosophy,"* New York: Harper.
2. A sarcastic quip at the time was that there were "wavicles."
3. de Broglie, L. (1953). *The Revolution in Physics,* New York: The Noonday Press.
4. Grad, H. (1961). "The Many Faces of Entropy," *Communications of Pure and Applied Mathematics,* **14**, 323–354.
5. Prigogine, I. (1980). *From Being to Becoming* (p. 51). New York: W. H. Freeman and Co.
6. Nernst, W. (1969). *The New Heat Theorem,* New York: Dover Publications.
7. The level of activity of particles is determined by their kinetic energy, which is $1/2 \, mV^2$, where $\{m\}$ is the particle mass and $\{V\}$ the particle velocity. Let us further recall our previous brief discussion of the degrees of freedom. A particle that can move in the x, y, and z directions has a translational kinetic energy of $3/2 \, kT$ where $\{k\}$ is the universal Boltzmann constant and $\{T\}$ is the absolute temperature. Because both relations define the kinetic energy of a particle we can set them equal to another. To be specific let us do this for an electron gas as in the solar wind, or in a fusion reactor. Introducing the mass of the electron $\{m_e\}$, we then find that 10,000° Kelvin is approximately equal to 1.29 ev. Although we are not accustomed to dealing with temperature in this manner when we listen to the evening weather report, it is a common way of comparing research devices.
8. Because irreversible processes cannot be retraced back to their original position without leaving any evidence, they have trajectories rather than neatly closed loops.
9. Heisenberg, W. (1958). *op. cit.*
10. Dyson, F. J. (1971). "Energy in the Universe," *Scientific American,* **255**(3), 51–59.
11. Cardwell, D. S. L. (1971). *From Watt to Clausius,* Ithaca, NY: Cornell Univ. Press.
12. See Stodola, A. (1945). *Steam and Gas Turbines,* New York: Peter Smith.
13. Weinberg, S. (1988). *The First Three Minutes,* New York: Basic Books.
14. Shapiro, A. H. (1953). *The Dynamics and Thermodynamics of Compressible Fluid Flow,* New York: The Ronald Press Company.
15. Tribus, M. (1961). *Thermostatics and Thermodynamics,* Princeton. NJ: D. Van Nostrand Company.
16. Prigogine, I., and Stengers, I. (1984). *Order out of Chaos,* Toronto: Bantam Books.
17. Onsager, L. (1931). *Physical Review* **37**, pp. 405 ff., and **38**, pp. 2265 ff.
18. Prigogine, I. (1947). *Etude Thermodynamique des Phénomènes irréversibles,* Paris: Dunod Editeurs.
19. Prigogine, I. (1980). *From Being to Becoming,* New York: W. H. Freeman and Co.
20. Glansdorff, P., and Prigogine, I. (1971). *Thermodynamics of Structure, Stability, and Fluctuations,* New York: John Wiley & Sons.
21. Nicolis, G., and Prigogine, I. (1977). *Self-Organization in Nonequilibrium Systems,* New York: John Wiley & Sons.
22. Prigogine, I. (1947), *op. cit.*
23. Çambel, A. B. (1984). "A Synergistic Approach to Energy-Oriented Models of Socio-Economic–Technological Problems," *Synergetics—From Microscopic to Macroscopic Order* (pp. 183–196). E. Frehland ed., Berlin: Springer-Verlag.
24. Prigogine, I., and Stengers, I. (1984). *op. cit.*
25. Nicolis, G., and Prigogine, I. (1977). *op. cit.*
26. Cramer, F. (1984). "Death—From Microscopic to Macroscopic Order," *Synergetics—From Microscopic to Macroscopic Order* (pp. 220–228). E. Frehland, ed., Berlin: Springer-Verlag.

27. Yourgrau, W., van der Merwe, A., and Raw, G. (1966). *Treatise on Irreversible and Statistical Thermophysics*, New York: Dover Publications.

28. Dyson, F. J. (1971). *op. cit.*

29. Schrödinger, E. (1943). *What if Life?*, Cambridge: Cambridge Univ. Press,

30. Brillouin, L. (1964). *Scientific Uncertainty and Information*, New York: Academic Press.

31. Wiener, N. (1948). *Cybernetics*, New York: John Wiley & Sons.

32. Wiener, N. (1968). "Cybernetics in History," in *Modern Systems Research for the Behavioral Scientist*, (pp. 31–36), W. Buckley, ed., Chicago: Aldine Publishing Co.

33. Shannon, C. E. (1948). "A Mathematical Theory of Communication," *The Bell System Journal*, **XXVII**, 379–423 and 623–656.

34. The aforementioned papers may be difficult to locate. Fortunately, Shannon's papers have been reprinted together with an expository chapter by Warren Weaver. See Shannon, C. E. and Weaver, W. (1963). *The Mathematical Theory of Communication*, Urbana, IL: Univ. of Illinois Press.

35. If the natural logarithm to the base {e} is used, the unit of information is called the "nat," while if binary counting is used, the unit of information is the "bit." The conversion factor is 1 nat = 1.443 bits. I am indebted to my former graduate student, Sheila R. Moore, for calling these units to my attention.

36. Andrews, E. C. (1975). *Equilibrium Statistical Mechanics*, New York: John Wiley & Sons.

37. Tribus, M., and McIrvine, E. C. (1971), "Energy and Information," *Scientific American*, **224**(3) 179–190.

38. This story was told to me by the late Theodore von Kármán, in the middle of Times Square in New York while automobiles were whizzing by!

39. Eckmann, J.-P., and Ruelle, D. (1985). "Ergodic Theory of Chaos and Strange Attractors," *Reviews of Modern Physics*, **57**(3), 617–656.

40. Lebowitz, J. L., and Penrose, O. (1973). "Modern Ergodic Theory," *Physics Today*, February,. pp. 23–29.

41. Ornstein, D. D. (1989). "Ergodic Theory, Randomness, and 'Chaos'," *SCIENCE*, **243**, 13 January, pp. 182–187.

42. Petersen, K. (1983). *Ergodic Theory*, Cambridge: Cambridge Univ. Press.

43. E. Whittaker, (1958). *From Euclid to Eddington*, New York: Dover Publications.

44. Bergson, H. (1950). *Time and Free Will*, (Translated by F.L. Pogson), London: G. Allen & Company.

45. Brillouin, L. *op. cit.*

46. Eddington, A. S. (1929). *The Nature of the Physical World*, New York: The Macmillan Co.

47. Khayyam, O. (circa twelfth century). *Rubaiyat of Omar Khayyam*, (Translated by E. Fitzgerald), New York: Hartsdale House.

48. Hawking, S. W. (1988), *A Brief History of Time*, Toronto: Bantam Books.

49. Murphy, R. E. (1965). *Adaptive Processes in Economic Systems*, New York: Academic Press.

50. Fraser, J. T. (1988). *Time the Familiar Stranger*, Redmond, WA: Tempus Books of Microsoft Press.

51. Boslough, J. (1990). "The Enigma of Time," *National Geographic*, **177**, 109–132.

52. Coveney, P., and Highfield, R. (1990). *The Arrow of Time*, New York: Fawcett Columbine.

53. Schuster, H. G. (1988). *Deterministic Chaos*, Weinheim, Germany: VCH Verlagsgesellschaft.

54. Baker, G. L., and Gollub, J. P. (1990). *Chaotic Dynamics*, Cambridge: Cambridge Univ. Press.

55. Rényi, A. (1955). "On a New Axiomatic Theory of Probability," *Acta Mathematica Hungarica,* **6,** 285–335.

56. Rényi, A. (1960). "On Measures of Entropy and Information," *Proc. 4th Berkeley Symposium on Mathematics, Statistics, and Probability,* pp. 547–561.

57. Farmer, J. D. (1982). "Information Dimension and the Probabilistic Structure of Chaos," *Zeitschrift für Naturforschung,* **37a,** 1304–1325.

58. Grassberger, P. (1983). "Generalized Dimensions of Strange Attractors," *Physics Letters,* **97A**(6), 227–230.

59. Schroeder, M. (1991). *Fractals, Chaos, Power Laws,* New York: W. H. Freeman and Co.

9

DIMENSIONS AND SCALING

INTRODUCTION

Shape and size are important whether we are buying clothing or a vacant lot. Not all shapes are easily described in terms of rectangles, circles, spheres, etc. For example, we considered earlier the strange attractors associated with chaos which have irregular shapes and hence do not fit conventional patterns. Still it becomes necessary to classify them. A convenient approach is to speak of their characteristic *dimension(s)*. These are not necessarily the length, the width, and the height that we are accustomed to using. For example, we have already dealt with two dimensions: the Lyapunov dimension and the entropy, both rather different types of dimensions from what we use in daily life. Irregular shapes, such as strange attractors, seashores, and the boundaries between countries, are not easily measured because they are *fractal*. In other words, they have non-integer dimensions, in contrast to the integer dimensions that we casually use each day.

The colloquial expression "sizing up something" is actually very far-reaching. It implies that we are making measurements in a variety of ways. These may include the height, girth, intelligence, personality, etc., of a person we meet. They may include the length, style, color, and speed of an automobile that passes us on the highway. Some automobiles resemble others so that although their physical size is different they look like scale images of one another. But so do the trees on the sides of the road. While some are taller than others, there is a commonality in that all have a trunk, branches, and twigs. And we start wondering: "Is there a difference between the similarity among the cars, and the similarity of the trees?" Yes, indeed there is. As a rule, natural objects are irregular, and they do not have the sleek lines, nor the distinct edges and corners, that manmade objects have. The French–American mathematician and dean of fractal geometry, Benoit Mandelbrot,[1] who is associated with both I.B.M. and Yale University, expressed it succinctly:

> Clouds are not spheres, mountains are not cones, coastlines are not circles, and bark is not smooth, nor does lightning travel in a straight line.

It is evident that if we are to deal with complex structures, we need a more general geometry, one that transcends the Euclidean geometry that we are taught in school. In this era, the popular name is "fractal geometry," the geometry where the dimensions are not integer numbers, but are fractional, such as 1.7 or 2.3. The term "fractal" derives from the Latin *fractus,* meaning "to break." It was introduced in 1975 by B. B. Mandelbrot. Actually, the limitations of Euclidean geometry were realized a long time ago. Mandelbrot[2] quotes Richard Bentley (1662–1742):

> We ought not . . . to believe that the banks of the ocean are really deformed, because they have not the form of a regular bulwark; nor that the mountains are out of shape, because they are not exact pyramids or cones; nor that the stars are unskillfully placed, because they are not all situated at uniform distance. These are not natural irregularities, but with respect to our fancies only; nor are they incommodious to the true uses of life and the designs of man's being on earth.

Natural forms exhibit an amazing structural integrity and orderliness. Cumulus clouds, a bed of mushrooms, and sand dunes all exemplify the orderliness of nature.

Nature also exhibits scaled self-similarity. As we remarked earlier we notice this in trees, whose trunks, branches, and twigs are self-similar. This may be seen in Fig. 1, which is the photograph of a tree by my home.

Figure 1. Tree in the winter in McLean, Virginia.

As any gardener knows, at the other end of the tree the root structure too is self-similar. Indeed, the importance of non-Euclidean geometry is gaining considerable importance in forestry and plant physiology, and has earned the name of *graftals*, and *Lindenmayer-* or *L-systems*.[3] Fractals and self-similarity abound otherwise in nature—our own circulatory and bronchial systems are typical. Figure 2 shows the structure of a Chinese cabbage leaf.

Figure 2. Chinese cabbage leaf.

DIMENSIONS

As we have noted, complex systems can exhibit similarity, and they can be scaled. Accordingly, we must establish their dimensions. For purposes of review the types of dimensions that are of primary interest to us are: (a) Euclidean dimensions, $\{D_E\}$, that we are well familiar with since our early school days; (b) topological dimensions, $\{D_T\}$, that derive by dividing one set by another. For example, volumes subsume areas, areas subsume lines, and lines subsume points. It follows that as a rule topological dimensions will be larger than their corresponding Euclidean dimension; (c) fractal dimensions, $\{D_F\}$, such as the Hausdorff–Besicovitch $\{D_{H-B}\}$, or capacity, $\{D_C\}$, dimensions. In general (Schaffer and Tidd[4]):

$$D_E > D_F > D_T \tag{1}$$

In an epochal paper, Farmer, Ott, and Yorke[5] outlined the dimension of chaotic attractors. They point out that there are three basic types of fractal dimensions. The "metric" dimensions include the Hausdorff–Besicovitch dimension and the capacity dimension. The metric dimensions of a strange attractor are independent of the frequency with which the attractor is visited. The second class of fractal dimensions is said to be "probabilistic" because it does depend on the frequency with which the attractor is visited. The information dimension that will be defined shortly is also a type of probabilistic dimension because it derives from Shannon's entropy.

HAUSDORFF–BESICOVITCH DIMENSION

The different aspects of dimensionality and scaling make it imperative to formulate a generalized explanation of what might be called the *characteristic dimension*. At present there is no one standard definition. Indeed there is a wide variety. Also, some dimensions have the same basic form even when approached from different viewpoints.

In the most fundamental way, the *size* or *effective dimension* {V} of a region increases in some way with its linear dimension {L}. We can express this generally as $V(L) \approx L^D$ where {D} is an effective dimension that may be an integer or noninteger number. If it is the latter we call it a *fractal dimension*. The above, while general in applicability, is not sufficiently specific for applications. Here I shall follow in the footsteps of both Schroeder[6] and Voss.[7]

We are accustomed to dealing with points, lines, areas, and volumes. Let us take a line segment 1 ft. long, and multiply it by 3. We get a line segment 3 ft. long. Next let us take a square, 1 ft. on each side. We have an area of 1 sq. ft. Now let us multiply each side by 3. We get an area that measures $3 \times 3 = 9$ sq. ft. Finally, let us take a cube, 1 ft. on each side. It has a volume of 1 ft³. Again let us multiply each side by 3. The volume of the cube will be $3 \times 3 \times 3 = 27$ ft.³ Whereas the dimension of the line changed linearly, the dimensions of the surface and the volume changed nonlinearly, to the second power for the area, and to the third power for the volume. The dimension of a line is 1, the dimension of an area is 2, and the dimension of a volume is 3.

What we did above was to "scale" the line, the area, and the volume by a *magnification factor* or *scale ratio* of {3}. This choice was completely arbitrary; it could have been 4, 5, or any other number, including numbers less than unity. It is just like zooming in or zooming out with a variable focal length lens. To generalize our experiment, let us denote by {r} the scale ratio and by {N} the number of parts. It follows that for a line, $Nr^1 = 1$; for a square, $Nr^2 = 1$; and for a cube, $Nr^3 = 1$. Let us in accordance with convention denote the exponent by {D}. In general, $Nr^D = 1$. Hence,

$$D = \frac{\log N}{\log (1/r)} \qquad (2)$$

Because no restrictions were specified, {D} is quite general. There is no inalienable requirement that dimensions must be presented in any particular way. What is important is that we can state the size or the *effective dimension* of the object under scrutiny.

The concept of a generalized dimension goes back to German mathematician Felix Hausdorff[8] (1868–1942). One denotes the total size of an object by $\{L\}$, and then either fills it or covers it with tiny hypercubes or hyperspheres, $\{N\}$ in number and each having size $\{r\}$. Hence we denote these by $\{N(r)\}$. The Hausdorff dimension is descriptive at any size including lines or curves having zero length, i.e., point sets. For this extreme condition one sets $L = 1$, and demands that $r \to 0$. Consequently,

$$D_{H-B} = \lim_{r \to 0} \frac{\log N(r)}{\log (1/r)} \tag{3}$$

It follows that for a point, the Euclidean as well as the H–B dimension are zero because $N(r) = 1$.

There are a variety of fractal dimensions. These include: Hausdorff or the Hausdorff–Besicovitch dimension, the capacity dimension, the fractal dimension, and the similarity dimension. If the result is a noninteger, we have a fractal dimension. For the nuances, please refer to Schroeder.[9]

For the uninitiated, using fractal dimensions can be puzzling because it complicates a matter that we thought we had understood for a long time. But then as geneticist J. B. S. Haldane remarked, "The universe is not only queerer than we supposed but queerer than we can suppose."

Table I lists a number of representative fractal dimensions in order to provide a feel of orders of magnitude. The first several items, namely, the Cantor set, the von Koch snowflake, the Sierpinsky triangle, and the Menger sponge, will be considered in the next chapter. A word of caution is in order. The values in Table 9.1 were collected from different sources, and hence their definitions are not consistent. Before using any of the numerical values the reader should consult the references cited.

For the convenience of the reader some typical fractal dimensions are shown in the table on the following page.

A table of the fractal dimensions of fragmented objects such as projectiles, asteroids, broken coal, etc., has been compiled by Professor D. L. Turcotte[25] of Cornell University. Such information can be useful in controlling crushing as well as fragmentation processes. The fractal dimensions of Earth topologies have been discussed by Huang and Turcotte[26] and compared with synthetic images.

In puzzling over dimensions it would be stimulating to read *Flatland* by Abbott.[27] Meanwhile let us practice with Eq. 2. Let us explore the consequences of scaling up by a factor of 2. In the case of the square we have $\log 4/\log 1 = \log 2^2/\log 2$, or, $2 \log 2/\log 2 = 2$. We knew all along that a surface is two-dimensional. Similarly, we can confirm that a volume is

Table 9.1. Selected Fractal Dimensions[10]

Item	Dimension	Source
Cantor set	0.63093	Formula for D_{H-B}
Koch snowflake	1.26186	Formula for D_{H-B}
Sierpinsky triangle	1.58496	Formula for D_{H-B}
Menger sponge	2.72683	Formula for D_{H-B}
Henon attractor	1.22	Schaffer–Tidd[11]
Lorentz attractor	2.06	Grassberger–Procaccia[12]
Logistic map	0.538	Grassberger–Procaccia
Rayleigh–Benard	2.8	Bergé–Pomeau–Vidal[13]
Taylor–Couette	3.5	Swinney[14]
Turbulence	2.1–4.1	Grassberger[15]
NH_4Cl crystal	1.67	Honjo–Ohta–Matsushita[16]
Clouds	2.35	Hentschel–Procaccia[17]
Geothermal rock	1.25–1.55	Cambel[18]
Woody plants	1.28–1.79	Morse et al.[19]
Sea anemone	1.6	Burrough[20]
Coast line	1.05–1.25	Mandelbrot/Richardson[21]
Speech forms	1.66	Pickover/Khorasani[22]
Capital markets	$2 \to 3$	Peters[23]
Turbulent flame	2.33	Kersten[24]

three-dimensional because doubling its dimensions will yield $\log 8/\log 2 = \log 2^3/\log 2 = 3 \log 2/\log 2 = 3$. This confirms our earlier assertion that in the limit the relation for the generalized dimension will yield results that are in conformity with Euclidean geometry. It is always comforting if a new explanation that covers a broader range than previously considered reduces to the original when limited to the previous range. It is reassuring to know that you can go home again if you accept previous paradigms.

EMBEDDING DIMENSION

In all walks of life we encounter time series that depict the variation of some variable over time. Typical examples are fluctuations of the stock market, cardiac rhythms, seismogram traces that follow the vibrations of the earth that are associated with earthquakes, and electroencephalograms that trace the electric impulses in the brain. Time series of innumerable types are the bread and butter of researchers. We shall consider their analyses in connection with the diagnostics of chaos.

It would certainly be helpful to understand the nature of these wiggles both to (a) discern what happened and (b) to predict what might happen in the near or distant future. One major problem in analyzing any type of data, particularly in a new field, is that we do not know which are the pertinent data that should be investigated. This problem is related to another source of uncertainty, namely the number of the applicable degrees of freedom. In some ways it is like the chicken and egg proposition. How can you measure something when you don't know what it is that you should be measuring?

In the previous pages we became acquainted with a number of dimensions that are useful in describing the level of complexity. What we did was particularly applicable to static conditions. However, time series depict events as they evolve. Fortunately, it is possible to describe the empirical dynamics geometrically.

The method that has proven to be useful originates from differential topology, a special branch of mathematics. Specifically, surfaces are said to be *homeomorphic* if they can be transformed into one another by a topological transformation. For example, a sphere and an ellipsoid are homeomorphic. On the other hand, a sphere and a torus are not homeomorphic. Think of a round balloon that you can squeeze as you try to visualize this, and then compare it with those long sausage-type balloons also sold at novelty shops. In other words, certain topological surfaces can be transformed into one another, while others cannot. The "homeomorphic" or "embedding" theorem states that if dim $X \leq N$, then X can be embedded in Y if $Y = 2N + 1$ (Jackson[28]). This means that we have affected a mapping into a different space, one which may have fewer dimensions. Having fewer dimensions is appealing because if we are dealing with an unexplored problem it is advantageous to have to deal with as few dimensions as possible. Stated differently, this means minimizing the degrees of freedom.

With the aforementioned philosophical background, we now outline the approach following Rasband.[29] For the type of system represented by the forced, damped pendulum, the phase space variables are x and $\dot{x}$ which are functions of time. From the calculus we recall that

$$\lim \Delta t \to 0 \, \frac{dx}{dt} = \frac{[x(t + \Delta t) - x(t)]}{\Delta t} \tag{4}$$

Thus knowing the trajectory of points $[x(t), x(t + \Delta t)]$ means knowing the trajectory $[x(t), \dot{x}(t)]$. However, we do not know how many points must be known in the phase trajectory $x(t) = [x_1(t), x_2(t) \ldots, x_n(t)]$. Accordingly, we construct an artificial or pseudophase space. We get a series of points like

$$\mathbf{x}(t) = [y(t), y(t + \Delta t) \ldots, y(t + m \Delta t)]$$

.

.

.

A *correlation dimension* is then calculated from these figures. Here, too, a judgment call must be made concerning the particular definition of the correlation dimension because once again there is no unanimous agreement. A generic definition given by Schroeder[30] is

$$D_C = \lim_{(r \to 0)} \frac{\log \sum p_i^2}{\log r} \tag{5}$$

where p_i are probabilities.

This equation has the advantage of being akin to the information dimension

$$D_I = \frac{N \log N}{\sum (1/r)} \tag{6}$$

The numerator suggests that in turn this equation is related to the entropy.

SCALING

One of the important characteristics of fractals is *self-similarity*. This means that a fractal image may be scaled up or down, i.e., the magnification with which it is observed may be increased or decreased, yet the basic shape of the fractal remains the same. This is a very important attribute and allows one to do practical things such as image compression. We start our discussion with a very mundane example. Specifically, let us attempt to determine on a road map the distance to be travelled for a vacation trip. Much depends on the scale size of the ruler and also the patience exercised in negotiating around the minuscule twists and turns. Two different persons using the same scale will probably arrive at different distances. Indeed it is likely that the same person using the same scale will measure different distances when the procedure is repeated at another time.

In an early paper Mandelbrot[31] asked: "How long is the coast of Britain?" It depends, because any measurement inextricably involves the object, the measuring scale, and the observer performing the measurement. Quite obviously the exact length of the British coastline would depend on how and by whom the measurement is performed. Does one clock the odometer on the dashboard of a Bentley travelling the road along the shoreline, or the pedometer of a hiker on the walking path immediately adjacent to the shoreline? How about an ant armed with a tiny measuring device? In any case, the smaller the scale, the closer it can follow the contour, and the more accurate the dimension will be. In all probability, the length of the contour will be longer if the scale is made smaller. In short, the accuracy of the measurement depends on the scale used. This could have serious implications whether it be the establishment of boundaries among nations, or the successful assembly of products from parts that were measured with different scales during manufacturing.

One of the earliest investigators to look into this problem was Quaker meteorologist Lewis F. Richardson,[32] who had a longstanding interest in the causes of international conflict. One of his investigations involved the length of national borders between contiguous countries. He described his procedure in the following words:

> As an explanation of how chance can arise in a world which he regarded as strictly deterministic, Henri Poincaré drew attention to the insignificant causes which produced very noticeable effects. Seacoasts provide an apt illustration. For the spike of the dividers may just miss, or just catch, a promonotory of land or the head of a loch. . . .

Evidently, Richardson was simmering on the edge of the concept of fractals. He made detailed measurements by walking a pair of dividers along the frontiers depicted on maps, and counted the number of the equal sides of polygons adjacent to the frontier. By making numerous measurements of maps, Richardson arrived at the following proportionality:

$$\Sigma L \propto L^{-D_R} \tag{7}$$

where $\{L\}$ is the length of a typical polygon edge on a boundary, $\{\Sigma L\}$ is the total length, and the exponent $\{D_R\}$ is called the *Richardson dimension*. For example, for the coast of England, which is ragged, he found $\{D_R\}$ to be about 0.25, while for South Africa $\{D_R\}$ is of the order of 0.02, indicating its relative smoothness. Contours that have the same value of $\{D_R\}$ have the same degree of raggedness; they enjoy scale invariance.

It has been shown by Mandelbrot that

$$D_{H-B} = D_R + 1 \tag{8}$$

Because scale invariance is a common occurrence in geological phenomena, Richardson's technique has been extended and is applied in geophysics and the earth sciences (Turcotte[33]).

Scaling is an obvious manifestation of complexity. If your hobby is miniature trains, you know that no matter how intricate your models, they cannot include all the details of the prototype locomotive that pulls the Amtrak Metroliner. Nor can you scale up indiscriminately that cute little cottage you fell in love with; the large structure may cave in under its own weight.

The limitations of scaling extend past romantic examples. Scaling affects major decisions on public policies. In the early days of the nuclear energy industry it was conjectured that the larger the power plant capacity, the lower would be its unit-installed capacity cost. This conjecture was stimulated by extrapolating linearly the cost of a few small nuclear power plants to large sizes. By doing so it was thought that the cost would be reduced very significantly. The results were so enticing that a high-level official was moved to testify that nuclear energy would be so inexpensive that ultimately its metering would become unnecessary. This was an absurd claim because in any growth situation nonlinearities eventually appear. Thus there developed a tendency to build larger and larger plants. Ultimately this became counterproductive technologically and economically. It must be noted that systematic efforts for understanding nonlinear problems started in World War II. In 1960 the United States Atomic Energy Commission published the pioneering book on nonlinear equations, written by Professor H. T. Davis[34] of Northwestern University.

So far we have concerned ourselves with geometric scaling. I shall now introduce the concept of *fractal time series,* or *biased random walks.* It is evident that it is instructive to compare the fractal dimensions of dynamic systems over periods of time. One approach is the so-called *rescaled range analysis* originated by H. E. Hurst[35] (1900–1978) and amplified by Hurst, Black, and Simaika.[36] Applications and extensions may be found in Feder,[37] and Schroeder.[38] It is interesting to note that Peters[39] applied this technique to the analysis of capital markets.

For many years Hurst studied water storage in reservoirs. A major application would be in the design of dams, another would be the overflowing or drying up of lakes around which cities are erected. The technique is also of interest for the optimum design of artificial lakes that are built for community solar ponds, or in which heat exchangers are immersed for process heating and/or cooling. Ideally such a reservoir would neither overflow nor empty. However, the water level would rise and fall with rainfall, by the flux of tributaries, and climactic effects. The range $\{R\}$ between the maximum and minimum water levels over the time period of $\{\tau\}$, usually given in years, may be written as follows:

$$R(\tau) = \max \ z(t, \tau) - \min \ z(t, \tau) \tag{9}$$

Hurst found that the ratio $\{R/S\}$ where $\{S\}$ is the statistical standard deviation is a useful scaling factor that he applied to many natural objects and events such as river discharges, rainfall, temperature, and tree rings. Peters[40] applied this "rescaled range analysis" in his analyses of the market place. The empirical equation that Hurst arrived at is the following:

$$\frac{R}{S} = \left(\frac{\tau}{2}\right)^{H} \tag{10}$$

where H is the *Hurst coefficient,* which varies with the situation. Tabulated values of $\{H\}$ may be found in Hurst *et al.*[41] Another expression for $\{R/S\}$ is given by Peters:

$$\frac{R}{S} = (a * n)^{H} \tag{11}$$

where $\{a\}$ is a constant and $\{n\}$ the number of observations.

In applications of chaos theory it is always useful to know the role of randomness. Peters[42] has expressed the impact of the present on the future as follows:

$$C = 2^{(2H-1)} - 1 \tag{12}$$

where $\{C\}$ is the correlation measure. He further differentiates among three values of $\{H\}$. When $H = 0.5$, we get $C = 0$, i.e., the events are random and uncorrelated. When $0 \leq H < 0.5$, reversals can occur and, in the vernacular, the situation is volatile. On the other hand, when $0.5 < H < 1.0$, the trend is persistent.

In ending this chapter it is interesting to ask if there is any relationhip between position scaling and time scaling. It turns out that there is (Feder[43]). Specifically,

$$D_{H-B} = 2 - H \tag{13}$$

We have been exposed to numerous dimensions and different ways of scaling. There is no one correct recipe of combining them, because we are still in the developmental stages of chaos theory. I find that every time I apply them I understand them better, even if the first attempt is unsuccessful.

Scientific details of fractal time may be found in the review paper[44] by M. F. Shlesinger, pioneer in complex sciences.

NOTES AND REFERENCES

1. Mandelbrot, B. B. (1983). *The Fractal Geometry of Nature*, (p. 1), San Francisco: W. H. Freeman and Co.
2. Mandelbrot, B. B. (1983). *op. cit*, p. 6.
3. Prusinkiewicx, P., and Hanan, J. (1989). *Lindenmayer Systems, Fractals, and Plants*, New York: Springer-Verlag.
4. Schaffer, W. M., and Tidd, C. W. (1990). *Chaos in the Classroom: II Fractals and Julia Sets*, Leesburg, VA: Campus Technology.
5. Farmer, J. D., Ott, E., and Yorke, J. A. (1983). "The Dimension of Chaotic Attractors," *Physica*, **7D**, 153–180.
6. Schroeder, M. (1991). *Fractals, Chaos, Power Laws*, New York: W. H. Freeman and Co.
7. Voss, R. F. (1988). "Fractals in Nature: From Characterization to Simulation," in *Fractal Images*, M. F. Barnsley, R. L. Devaney, B. B. Mandelbrot, H.-O. Peitgen, D. Saupe, and R. V. Voss, eds. (pp. 21–70). New York: Springer-Verlag.
8. Hausdorff, F. (1919). "Dimension und äusseres Mass," *Mathematische Annalen*, **79**, 157–179.
9. Schroeder, M. (1991). *op. cit.*
10. The reader is cautioned that there may be inconsistencies in this table because the entries have different origins.
11. Schaffer, W. M., and Tidd, C. W. (1990). *op. cit.*
12. Grassberger, P., and Procaccia, I. (1984). "Dimensions and Entropies of Strange Attractors from a Fluctuating Dynamics Approach," *Physica*, **13D**, 34–54.
13. Bergé, P., Pomeau, Y., and Vidal, C. (1984). *Order within Chaos*, New York: John Wiley & Sons.
14. As per Heppenheimer, T. A. (1986). "Routes to Chaos," *Mosaic*, **17**, Summer, 2–13.
15. Grassberger, P. (1986). "Estimating the Fractal Dimensions and Entropies of Strange Attractors," (pp. 291–311), in *Chaos*, A. V. Holden, ed. Princeton, NJ: Princeton Univ. Press.
16. Honjo, H., Ohta, S., and Matsushita, M. (1986). "Irregular Fractal-Like Growth of Ammonium Chloride," *Journal of Physical Society of Japan*, **55**, 2487.
17. Hentschel, H. G. E., and Procaccia, I. (1983). "The Infinite Number of Generalized Dimensions of Fractals and Strange Attractors," *Physica*, **17 D**, 435–444.
18. Çambel, A. B. (1990). Preliminary Report to Sandia National Laboratories.
19. Morse, D. R., Lawton, J. H., Dodson, M. M., and Williamson, M.H. (1985). "Fractal Dimension of Anthropod Body Lengths," *Nature*, **315**, 731–733.
20. Burrough, P. A. (1981). "Fractal Dimensions of Landscapes and Other Environmental Data," *Nature*, **295**, 240–242.
21. Mandelbrot, B. B. (1983). *The Fractal Geometry of Nature*, San Francisco: W. H. Freeman and Company.
22. Pickover, C. A., and Khorasani, A. (1986). "Fractal Characterization of Speech Waveform Graphs," *Computers and Graphics*, **10**, 51–61.
23. Peters, E. E. (1991). *Chaos and Order in the Capital Markets*, New York: John Wiley & Sons.

24. Kersten, A. R. (1988). "Fractal Dimension of Turbulent Premixed Flames," *Combustion Science and Technology*, **60**, 441–445.
25. Turcotte, D. L. (1986). "Fractals and Fragmentation," *Journal of Geophysical Research*, **91**, 1921–1926.
26. Huang, J., and Turcotte, D. L. (1989). "Fractal Mapping of Digitized Images: Application to the Topography of Arizona and Comparisons with Synthetic Images," *Journal of Geophysical Research*, **94**(B6), 7491–7495.
27. Abbott, E. A (1952). *Flatland*, New York: Dover Publications.
28. Jackson, E. A. (1989). *Perspectives of Nonlinear Dynamics*, Cambridge: Cambridge Univ. Press.
29. Rasband, S. N. (1990). *Chaotic Dynamics of Nonlinear Systems*, New York: John Wiley & Sons.
30. Schroeder, M. (1991). *op. cit.*
31. Mandelbrot, B. B. (1967). "How Long Is the Coast of Britain? Statistical Self-Similarity and Fractional Dimension," *Science*, **155**, 636–638.
32. Richardson, L. F. (1960). "The Problem of Contiguity: An Appendix to Statistics of Deadly Quarrels," *General Systems—Yearbook of the Society for General Systems Research.* **V** (pp. 139–187), L. von Bertalanffy and A. Rapoport, eds., Ann Arbor, MI
33. Turcotte, L. D. (1991). "Fractals in Geology; What Are They and What are They Good For?" *GSA Today*, January, pp. 1, 3–4.
34. Davis, H. T. (1962). *Introduction to Nonlinear Differential and Integral Equations*, New York: Dover Publications.
35. Hurst, H. E. (1951). "Long-term Storage Capacity of Reservoirs," *Transactions of the American Society of Civil Engineers*, **116**, 770–808.
36. Hurst, H. E., Black, R. P., and Simaika, Y. M. (1965). *Long Term Storage: An Experimental Study*, London: Constable.
37. Feder, J. (1988). *Fractals*, New York: Plenum Press.
38. Schroeder, M. (1991). *Fractals, Chaos, Power Laws*, London: W. H. Freeman and Co.
39. Peters, E. E. (1991). *op. cit.*
40. Peters, E. E. (1991). *op. cit.*
41. Hurst, H.E., Black, R.P., and Simaika, Y. M. (1965). *op. cit.*
42. Peters, E. E. (1991). *op. cit.*
43. Feder, J. (1988). *op. cit.*
44. Shlesinger, M. F. (1988). "Fractal Time in Condensing Matter," *Annual Reviews of Physical Chemistry*, **39**, 269–290.

CHAPTER **10**

GALLERY OF MONSTERS

INTRODUCTION

In contrast to the docility of figures that comply with Euclidean geometry and its rules, a great many shapes follow the beats of another drum. Typically, such curves have ambiguous dimensions, they are not differentiable, they have no tangent and slope. Even the great mathematical innovator Poincaré was moved to call them "a gallery of monsters" (Briggs and Peat[1]). However, Boltzmann was excited about nondifferentiable functions and wrote that statistical mechanics required them. The French physical chemist and 1926 Nobel laureate in physics, Jean Baptiste Perrin (1870–1942), who contributed so much to Brownian motion, said:

> . . . curves that have no tangents are the rule, and regular curves, such as the circle, are interesting but quite special (Schroeder[2]).

Irregular, contorted geometric figures used to be looked upon as aberrations: They were called "pathological figures." Actually, such shapes are not aberrations nor exceptions: They are quite common. This realization has resulted in the formulation of new concepts in geometry that come under the rubric of *fractal geometry* because they have noninteger dimensions. This characteristic sets fractal geometry apart from Euclidean geometry with its integer dimensions. Thus in Euclidean geometry a point has zero dimensions, a line has one dimension, an area has two dimensions, and a volume has three dimensions, all integers. In contrast, fractal dimensions may assume values such as 0.538 for the discrete logistic curve, 1.6 for the sea anemone, 2.35 for clouds, and 3.5 for Taylor–Couette flow. Such fractional values should not surprise us; in the previous chapter we became acquainted with a number of nontraditional dimensions. In this chapter we shall become acquainted with the geometry of fractal shapes.

Fractal geometry is important in studying complexity for several reasons: First, almost all natural objects have irregular shapes and hence require more general dimensions than Euclidean geometry allows. Second, many complex systems are chaotic, and with these are associated strange attractors. These attractors have bizarre shapes and hence they cannot be described in terms of integer dimensions. Third, dynamic systems can be represented by time series and their dimensions are important if we are to characterize them. Fourth, fractals have an important characteristic: They tend to be self-similar. Thus one can scale fractals. We can compress or expand them without changing their basic shape. Closer inspection reveals more structural detail. Scaling allows us to replicate natural settings or complex shapes.

Nature abounds with fractals shapes, and these are attractively documented by photographer Eliot Porter and best-selling author James Gleick,[3] and by physicist and photographer Michael McGuire.[4] The joining of mathematics and exquisite computer-generated images may be found in the volumes by Peitgen and Richter,[5] Peitgen and Saupe,[6] and Guyon and Stanley.[7] The lavish color portraits have been themes for museum exhibits worldwide and have been used to illustrate calendars. With appropriate computer expertise one can generate fascinating images for the eyes to feast on. Amazing images can be produced by recursive figures, called "tiling," or "tessellation." The recursive nature of such patterns offers endless opportunities for letting the imagination roam wide and afar, as has been shown by I.B.M. biochemist Clifford Pickover.[8,9]

BACKGROUND

Dutch artist M. C. Escher (1898–1972) is particularly well recognized for his recursive patterns and counterintuitive scaling. His interdisciplinary outlook is the subject of considerable debate, and an international congress[10] was specially organized to discuss his art work and related science. His mastery in exploiting projective geometry is unique, particularly when one considers that he did not have computers at his service. The reader who wishes to be introduced to computer-oriented projective geometry will most likely benefit from the comprehensive volume by F. S. Hill, Jr.[11] However, there is a long history that leads to our present understanding.

In the early 1820s, Russian mathematician Nicolai Ivanovich Lobachevsky[12] (1793–1856) and Hungarian mathematician Janos Bolyai (1802–1860) independently developed the concepts of projective geometry. English translations of both Bolyai's discourse on absolute space and of Lobachevsky's theory of parallels may be found in Bonola.[13] According to Morris Kline,[14] Karl Friedrich Gauss (1777–1855) probably preceded them and in an unpublished work pointed out contradictions in Euclidean geometry. Specifically, the disconcerting point is that Euclid's fifth axiom, the parallel postulate, is inconsistent with the first four. It is conceivable that it was Gauss's interest in Renaissance paintings that made him aware of projective geometry, which led him to explore Euclidean geometry. Whatever the stimulus, Gauss's views about geometries other than Euclidean were antithetical to the prevailing Kantian views (Banchoff[15]). Certain issues remain unresolved to this day.

Another slice of this history traces to mathematician Karl Wilhelm Weierstrass (1815–1897), who in 1861 showed that certain continuous functions may have no tangent and may not be differentiable. This was an insult to calculus, because it challenged the concepts of continuity, slopes, and velocities. Mathematicians of the day rationalized it by arguing that this occurred in functions that were so complicated that they were abnormal and hence did not deserve attention. Little did they perceive.

Still another shock arrived in 1890 from Giuseppe Peano (1858–1932) when he described *space filling curves*, now named after him. Such curves are due to a moving point that in finite time passes through every point inside a square and eventually fills it entirely. A year after Peano's publication, the great German mathematician David Hilbert (1862–1943) presented his version, and there are others. For elaborations see Gardner,[16] Kline,[17] Newman,[18] and Peitgen, Jürgens, and Saupe.[19] There are different ways of constructing Peano figures. In Fig. 1 are shown the first three steps of Hilbert's construction.

Peano and Hilbert were interested in mathematical ambiguity. We know that in Euclidean geometry a line has one dimension. Accordingly, it was customary to consider a curve, i.e., a bent line, to be one-dimensional, too.

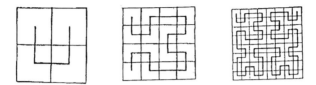

Figure 1. Early steps in Hilbert's construction of the Peano curve.

What Peano did was to show unequivocally that by moving a point (zero-dimension), figures that cover an entire plane can be generated in finite time and without the curve segments ever crossing themselves. In other words, 0-D is transformed into 2-D. "What are we dealing with," Peano asked, "a point, a line, or an area?" The present interest is due to the observation that in soils, plants, and living organisms fluids can reach the points they must because of such elaborate duct systems. It has been shown by Goldberger, Rigney, and West[20] that bronchia and blood vessels are fractal. The lungs and kidneys are three-dimensional Peano structures.

The geometric construction shown in Fig. 1 starts with a square, which is divided into four squares. Their center points are connected. Next, one again starts with the same empty square, divides it into 16 squares and again connects all of the centers. In subsequent steps the square is divided into 64, 256, 1,024, and 4,096 squares. One can continue doing this until all points within the square are connected, and the entire area is covered. It must noted that the line does not cross itself at any time.

There are several basic configurations that are traditionally used in introducing the subject of fractal geometry. These are the Cantor set, the von Koch snowflake, the Sierpinsky triangle or gasket, and the Menger sponge. For erudite treatments of fractals see the treatises by Mandelbrot,[21] Feder,[22] Falconer,[23] Peitgen, Jürgens, and Saupe.[24]

Cantor Set

Let us first consider the Cantor set. As shown in Fig. 2, we start with a straight line that may have any length. For convenience we consider a line of unit length.

We remove the middle section, but leave the two remaining sections, which are each one-third of the length. We next divide each of these lines into three, and again remove the middle third. We continue this iterative process. It becomes apparent at once that as the number of iterations

Figure 2. Cantor set.

increases, i.e., tends to infinity, the number of remaining sections will tend to infinity, while their respective lengths will tend to zero. In other words, we get an infinite number of points that we know to have zero dimension. Mandelbrot has considered the case of a three-dimensional Cantor set where one starts with a solid bar. After the iterations this culminates in "Cantor dust." This conceptualization could have applications in a variety of fields.

One can expect the dimension of the Cantor set to be someplace between one at the beginning and zero at the conclusion of the iterative process. This can be easily confirmed by applying the dimension equation for $\{D\}$ from the previous chapter. At the onset, $N = 1$ and $r = 1$. After the first excision, $N = 2$ and $r = 1/3$. After the second excision, $N = 4$ and $r = 1/9$, while after the third excision, $N = 8$ and $r = 1/27$. At the nth stage, $N = 2^n$ and $r = (1/3)^n$. It follows that $D = \log 2^n / \log 3^n = 0.630929753$. The Cantor set has a dimension that is less than unity.

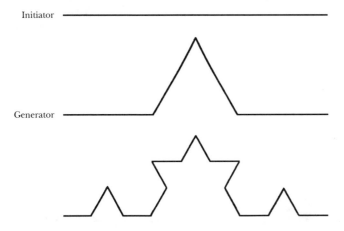

Figure 3. Construction of von Koch curve.

Figure 4. Construction of von Koch snowflake.

von Koch Snowflake

One of the earliest pathological figures is due to Swedish mathematician Helge von Koch (1904). It has two versions—one is open (Fig. 3) and the other is closed (Fig. 4). The latter is called the von Koch snowflake because of its resemblance to ice crystals.

In the open configuration, one starts with a straight line and, as in the case of the Cantor set, divides it into three. The middle section is excised, but instead of leaving an empty space, one erects a tent that has sides of the same length as the excised segment. One then proceeds to make incisions into all segments, including the tent, and erects a new tent each time.

In the case of the closed version of the von Koch construction, one starts with an equilateral triangle, and the middle third of each side is cut out. Instead of leaving it empty we erect a tent that has sides equal in length to the cutout. Next we cut out one-third of each side of all the tents, and erect tents on each opening. We continue with this iteration process. It follows that for the von Koch snowflake, $D = \log 4/\log 3 = 1.261859507\ldots$, which is more than a line, but less than a surface. We note that as we iterate, the length of the perimeter increases. Specifically, after $\{n\}$ iterations the perimeter will increase by a factor of $(4/3)^n$.

Sierpinsky Triangle

Another pathological figure is the Sierpinsky triangle, shown in Fig. 5.

Figure 5. Construction of Sierpinsky triangle.

In constructing the Sierpinsky triangle one again starts with an equilateral triangle, but instead of erecting tents, one cuts out equilateral triangles. Initially there is one blackened triangle; in the second step there are three, and in the third step there are nine. Each one is a scaled-down replica of the triangles in the previous step. These steps can be continued. The Sierpinsky triangle has a dimension of 1.5849625. The reader should ponder why this is higher than that of the von Koch snowflake, which also starts with an equilateral triangle.

Menger Sponge

For the Menger sponge (not shown here), which is a three-dimensional figure, $D = 2.726833027$, not quite a complete volume.

A mundane analog is a cord of firewood which measures 4 ft. by 4 ft. by 8 ft., but which does not have an effective volume of 128 ft.^{3} ($4 \times 4 \times 8 = 128$) because there are voids among the logs. It turns out statistically that in actuality the effective volume of a cord of firewood is only about 80% of the nominal volume, e.g., 102.4 ft.^{3} It is not a solid volume. Sponges as well as organs like lungs are Menger figures.

Determining fractal dimensions can be confounding. Sometimes it is helpful to look at the complex shape by identifying and differentiating between *initiators* and *generators*. Reconsider the von Koch snowflake. The equilateral triangle is the initiator, and the tents are the generators. This concept is helpful when it is not at once known what {N} and what {r} to use. A simple example will demonstrate one approach.

Example: Compare the dimensions of an open and a closed symmetric triangle as shown in Fig. 6.

Solution Please note that the closed symmetric figure never crosses itself. In both cases the initiators have N = 3. However, for the open triangle the generator has N = 4, and for the closed symmetric triangle the generator has N = 7. Hence the dimensions are respectively:

Figure 6. Open and closed symmetric triangles.

$$D = \frac{\log 4}{\log 3} = 1.27$$

and

$$D = \frac{\log 7}{\log 3} = 1.77$$

JULIA AND MANDELBROT SETS

The interest in nonlinear problems and the availability of computers have given great impetus to iterative computations. One of the oldest techniques of numerical analysis is "Newton's method," which is mentioned in his *Principia*. In order to find the roots of an algebraic equation, one guesses what the roots may be and solves the equation. The answer is then reintroduced into the equation, and another answer is obtained. One patiently goes on in this recursive manner until one is satisfied in being close enough to the roots. There are a number of variations of Newton's method, and in 1879 it was extended by Lord A. Cayley to complex polynomials. We present the following form:

$$x_{n+1} = x_n - \frac{f(x_n)}{f'(x_n)} \tag{3}$$

While Newton's method is powerful, it remained in relative obscurity until the recent interest in fractals. Major problems were worked out and the method was thoroughly brought up to date by two mathematicians, Adrien Douady of France, and John H. Hubbard of the U.S.[25,26]

So far we have used the coordinate plane and the phase plane. In both the axes are real numbers. Now we shall use the *complex plane*, wherein we shall use the horizontal axis for the real numbers and the vertical axis for the imaginary numbers. Both axes will extend from the positive to the negative regimes. By means of iterative computations performed in the complex plane some of the most eye-catching figures are obtained. One that is particularly well known is the *Mandelbrot set*, which is shown in Fig. 7. The computations were performed using the software program by Schaffer and Tidd.[27]

Here we shall concentrate on the two that are preeminent: the Julia set and the Mandelbrot set. We have already seen the latter in Fig. 7, and we now present a Julia set in Fig. 8, which was also generated with the Schaffer–Tidd software program.

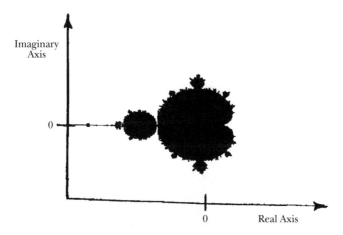

Figure 7. Mandelbrot set located in the complex plane. (Image of Mandelbrot set was generated with the W. M. Schaffer and C. W. Tidd computer program.)

We shall see shortly that the Julia set may assume different configurations, and that the one in Fig. 8 is only one of many.

It is appropriate to make two remarks. First, for convenience, and in accordance with common custom, I shall no longer show the axes of the complex plane. Second, to mimic color portraits I shall at times use black

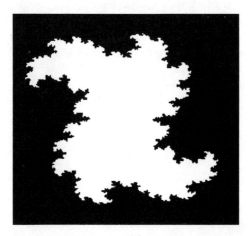

Figure 8. Julia set. (Image was generated with the W. M. Schaffer and C. W. Tidd computer software program.)

on white, and at times white on black images using the Raindrop[28] utility. This does not compromise the mathematics of the set being generated, but it does deprive one of identifying certain transitions, because color does help in determining the rate at which itineraries approach attractors. Multicolor renditions become a matter of choosing the palette arrangement and the configuration of the image grabber. For hard copy printing, the film and the optical filters also are important.

Complex numbers do not come as naturally as counting digits on one's fingers or toes. They were introduced in connection with the solution of quadratic algebraic equations. As simple a quadratic equation as one can think of is $x^2 - 1 = 0$. To solve it we simply transpose, so that $x^2 = 1$, and take the square root, i.e., $x = \sqrt{1}$. It immediately follows that $\{x\}$ has two solutions or roots, i.e., ± 1. Both +1 and −1 are *real numbers*. These are the numbers we use daily, and they are points on a line. The set of real numbers is denoted by $\{\mathbf{R}\}$. Real numbers may be positive or negative, integer or noninteger, rational or irrational.

Next, let us consider a slightly different quadratic equation, namely, $x^2 + 1 = 0$. When we attempt to solve this we find that $x^2 = -1$. Taking the square root yields $x = \sqrt{-1}$, which indicates that there are no real number solutions. To provide for this situation mathematicians defined *imaginary numbers*, specifically, the unit imaginary number $i^2 = -1$, or $i = \sqrt{-1}$. A complex number is then the sum of a real and an imaginary number. For example, the complex number x can be written $x = a + bi$ where $\{a\}$ and $\{b\}$ are real numbers, but $\{bi\}$ is an imaginary number because we multiplied the real number $\{b\}$ by the unit imaginary number $\{i\}$. The complex number x has two parts, the real part, $\{a\}$, and the complex part, $\{bi\}$.

Two French mathematicians, Pierre Fatou (1878–1929) and Gaston Julia (1893–1978), contributed in a major way to the iteration of complex functions and showed that the boundary points form sets. One such set named after Julia was shown in Fig. 8. The Julia set was extended by Benoit Mandelbrot, and the famous set is named after him.

The Mandelbrot set is expressed as follows:

$$x_{n+1} \Rightarrow x_n^2 + C \tag{4}$$

where $\{C\}$ represents complex numbers. In turn, the Julia set is expressed as:

$$x_{n+1} \Rightarrow x_n^2 + \mu \tag{5}$$

Here $\{\mu\}$ is a complex constant.

In the Julia set we let $\{x\}$ vary across the complex plane, and keep $\{\mu\}$ constant. In the Mandelbrot set, we let $\{C\}$ vary across the plane.

Regarding the iteration, we have two basic choices: We can stop iterating when the output exceeds a given value of $|x|$, commonly 2. Fractals are mathematical regimes, and we size them from the origin, $(0, 0i)$. Alternatively, we can stop the iteration when it reaches a specified number.

The number of iterations is crucial because it determines the detail that is revealed. See Fig. 9.

The two images in Fig. 9 were generated with the Sintar Software program KaleidoScope.[29] In the upper image the number of iterations was only 100, while in the lower figure the number of iterations was 10,000. The algorithm was the same in both cases. The difference in detail is evident.

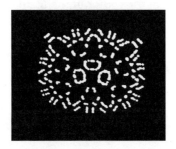

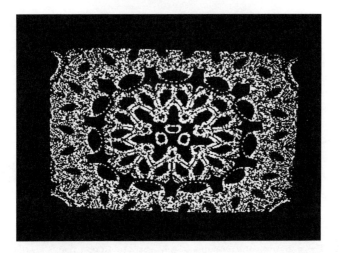

Figure 9. Upper image generated after 100 iterations, lower image after 10,000 iterations. (Images generated with kaleidoScope computer program by Sintar.)

Fractals are self-similar; they exhibit scale invariance. In other words, when we magnify repeatedly, the various images look the same regardless of size. This characteristic has implications in fields such as image compression, physiology, video–TV animation, and petroleum exploration. Fractals are made up of parts that are scale copies of the whole. In other words, there is replication. This becomes particularly pronounced in fractals such as the Mandelbrot set where as one keeps zooming in, clones keep on appearing. However, different Julia images appear when one gets close to different points along the rim of the Mandelbrot set. This is demonstrated in Fig. 10, which is a fractal montage.

In turn, Fig. 11 is a photo montage of the Julia sets at different points inside the buds of the Mandelbrot set. Here the nondescript image at the right side is the Julia set at the center of the main bud and corresponds to the point marked {+}. The figure at the top corresponds to the point marked

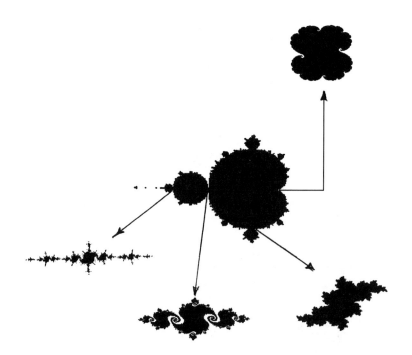

Figure 10. The different configurations of Julia sets at different points on the rim of the Mandelbrot set. (Images generated separetely with W. M. Schaffer and C. W. Tidd's program and then merged.)

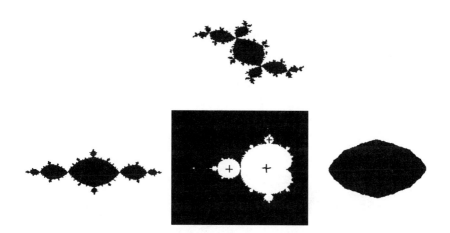

Figure 11. Photo montage of Julia sets corresponding to different points inside the buds of the Mandelbrot set. (Images generated with W.M. Schaffer and C. W. Tidd's software program, then merged).

{✛} at the center top bud, while the figure at the left of the Mandelbrot set is the Julia set for the left bud and the attractor has bifurcated to a 2-cycle (Schaffer and Tidd[30]).

The value of the complex number $C = a + bi$ determines whether or not the Julia set is within the Mandelbrot image. If the value is such that the point is inside, its location determines the shape of the Julia set, namely its complex dynamics. For example, in Fig. 11 when the point denoted by {✛} is in the main body, namely the cardiod, the Julia set, shown on the right side of the Mandelbrot set, is a deformed circle. If we move to the bud on the left of the main body, the cardiod, the corresponding Julia set that is depicted on the left of the Mandelbrot set is much more complex and indicates a bifurcation. This demonstrates the close relationship between the Mandelbrot set and the bifurcation map, and is illustrated in Fig. 12.

Thus the centerpiece of bifurcation theory and the centerpiece of fractal geometry—the Mandelbrot set—are actually united in geometric matrimony. In hindsight one could have expected this because Eq. 4 in this chapter and the discrete logistic equation we learned about previously have the same form; they differ only by the real and the complex terms.

In Fig. 12 the real axis is horizontal and runs through the antenna on the left, while the imaginary axis is vertically situated, running through

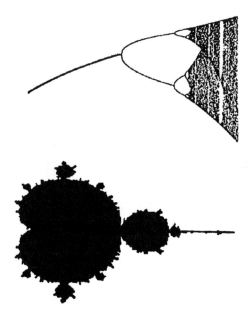

Figure 12. The correspondence of the Mandelbrot set and the bifurcation map.

two buds on the top and the bottom of the cardiod. The period-doubling scenario occurs along the real axis as different values are assigned to $\{C\}$. For any value of $\{C\}$ in the cardiod there will be a cycle of period-1 as demonstrated by the tail of the bifurcation diagram. In the bulb on the left of the cardiod, $C = 2$, which means that we have period doubling, as is evident in the bifurcation map. Further, a blip on the antenna may be noted. This corresponds to cycles of period-3 in the bifurcation map.

BARNSLEY'S CHAOS GAME

There are instances when it is useful to compress information for latter decompression according to a code. For example, a satellite rotating around a planet and taking photographs cannot easily transmit all of the data real-time. Professor Michael Barnsley of the Georgia Institute of Technology has developed a very ingenious method to codify fractals. The usefulness of this should become obvious if I inform you that half a page of text takes about 2,000 bits. In contrast, when I scan a fractal image covering about half a page of space, it takes about 30 times as many bits. Obviously, it will take a while to transmit this information. We need to do something about this. This is exactly what Barnsley[31] did when he developed "Iterated Function Systems" (IFS) with the aid of his "Chaos Game Algorithm." An IFS code is a formula for a fractal. It can be used to advantage when we wish to describe, classify, or communicate fractals to others. When the fractal changes, its IFS is changed; if the IFS is modified, the fractal takes on a different configuration. In applying IFSs, one uses *affine transformations*, which act on the Euclidean plane.

An affine transformation moves points in Euclidean space to another location by means of the following formula (Barnsley[32]):

$$W(x_{1,2}) = ax_1 + bx_2 + e(cx_1 + dx_2 + f) \qquad (6)$$

where a, b, c, d, e, and f are real numbers. An IFS is a collection of affine transformations that contains a finite number of W_n. In designing fractals

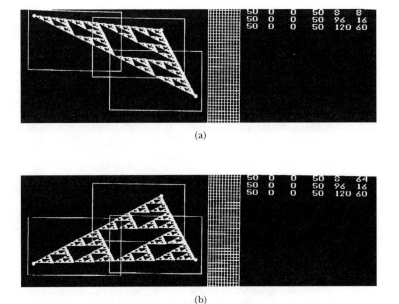

Figure 13. Rearrangement of Sierpinsky triangle. (Performed with M. Barnsley's software program.)

with the Barnsley program one picks images of attractors in the program library. Pieces of the attractor are surrounded by polygons. For each fractal image there is a corresponding code. One selects an active polygon and its corresponding active row in the code. With this very terse summation of Barnsley's approach let us perform some demonstrations. In Figs. 13a and 13b we see how the Sierpinsky triangle is moved about.

In each case the left window shows the attractor, and the right window shows the code. Note the coefficients of the codes before and after the move.

In Figs. 14a and 14b we see how a rectangular attractor picked from the Barnsley fractal library can be rendered into a Sierpinsky triangle. Please note the difference in the codes. Originally, there were four rows for the rectangle. After the transformation the first row is zero-valued throughout. For all practical purposes it has three rows just as in the Sierpinsky triangles in Fig. 13. The more complicated the attractor, the more polygons must be used, and consequently the code will become much more elaborate.

One of the interesting points of affine transformations and IFS is that one changes the fractal image by changing its IFS code. In other words, one

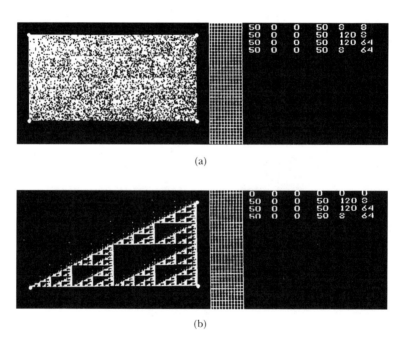

(a)

(b)

Figure 14. Transformation of attractor (Performed with M. Barnsley's software program).

can synthesize fractal images. Then knowing the code, one can replicate this elsewhere. Further, because the IFS codes are small they can be communicated easily by electronic mail. Another application is to match a fractal image in which we are interested, with one of the selections in the program's library and modify them sufficiently to arrive at a reasonable facsimile. The program will then give us the IFS code. This offers opportunities for energy resource explorers in the field. One of the first observations made is to study the rock formations in a terrain and then study their fissures and cracks. Equipped with any of many notepad-size computers, the explorers can match the geometry of the fissure, look up the corresponding IFS code, and communicate this via modem to the main laboratory, where the IFS code is translated back to the proper geometry of the fissure to look up the experience with such rocks. Çambel and Mock[33] have explored this for fissures in geothermal rock formations because the thermal effects of such rocks pose problems not commonly encountered in other rocks.

NOTES AND REFERENCES

1. Briggs, J. and Peat, F. D. (1989). *Turbulent Mirror*, New York: Harper & Row.
2. Schroeder, M. (1991). *Fractals, Chaos, Power Laws*, New York: W. H. Freeman and Co.
3. Porter, E., and Gleick, J. (1990). *Nature's Chaos*, New York: Viking.
4. McGuire, M. (1991). *An Eye for Fractals*, Redwood City, CA: Addison-Wesley Publishing Co.
5. Peitgen, H.-O., and Richter, P. H. (1986). *The Beauty of Fractals*, Berlin: Springer-Verlag.
6. Peitgen, H.-O., and Saupe. D. (1988). *The Science of Fractal Images*, New York: Springer-Verlag.
7. Guyon, E., and Stanley, H. E., eds. (1991). *Fractal Forms*, Amsterdam: Elsevier Science Publishers.
8. Pickover, C. A. (1990). *Computers, Pattern, Chaos and Beauty*, New York: St. Martin's Press.
9. Pickover, C. A. (1991). *Computers and Imagination*, New York: St. Martin's Press.
10. Coxeter, H. S. M., Emmer, M., Penrose, R., and Teuber, M. L. (1986). *M.C. Escher— Art and Science*, Amsterdam: North-Holland.
11. Hill, F. S. (1990). *Computer Graphics*, New York: Macmillan Publishing Co.
12. Some readers will recall the satirical song of the mathematician/composer/ singer/entertainer Tom Lehrer about Lobachevsky and the innuendos about scholarly success.
13. Bonola, R. (1952). *Non-Euclidean Geometry*, Translated by H. S. Carslaw, New York: Dover Publications, Inc.
14. Kline, M. (1972). *Mathematical Thought from Ancient to Modern Times*, Oxford: Oxford Univ. Press.
15. Banchoff, T. F. (1990). *Beyond the Third Dimension*, New York: Scientific American Library.
16. Gardner, M. (1989). *Penrose Tiles to Trapdoor Ciphers*, New York: W. H. Freeman and Co.
17. Kline, M. (1972). *op. cit.*
18. Newman, J. T. (1956). "The Crisis in Intuition," in *The World of Mathematics*, J. R. Newman, ed. **3** (pp. 1956–1976). New York: Simon and Schuster.

19. Peitgen, H.-O., Jürgens, H., and Saupe, D. (1992). *Fractals for the Classroom*, New York: Springer-Verlag.

20. Goldberger, A. L., Rigney, D. R., and West, B. (1990). "Chaos and Fractals in Human Physiology," *Scientific American*, **262**(2), 43-49.

21. Mandelbrot, B. B. (1977). *The Fractal Geometry of Nature*, San Francisco: W. H. Freeman and Co.

22. Feder, J. (1988). *Fractals*, New York: Plenum Press.

23. Falconer, K. (1990). *Fractal Geometry*, Chichester: John Wiley & Sons.

24. Peitgen, H.-O., Jürgens, H., and Saupe, D. (1992). *op. cit.*

25. Peitgen, H.-O. ed. (1989). *Newton's Method and Dynamical Systems*, Dordrecht, Netherlands: Kluwer Academic Publishers.

26. Douady, A. (1986). "Julia Sets and the Mandelbrot Set," in *The Beauty of Fractals*, (pp. 161–173), H.-O. Peitgen and P. H. Richter, eds. Berlin: Springer-Verlag.

27. Schaffer, W. M., and Tidd, C. W. (1990). *Chaos in the Classroom: II, Fractals and Julia Sets*, Leesburg, VA: Campus Technology.

28. This image inversion was accomplished with the print screen utility Raindrop, *Eclectic Systems, Springfield, VA* (1989).

29. Bolme, M. W. (1987). *Kaleidoscope*, Bellevue, WA: Sintar Software.

30. Schaffer, W. M., and Tidd, C. W. (1990). *op. cit.*

31. Barnsley, M. (1988). *Fractals Everywhere*, Cambridge, MA: Academic Press.

32. Barnsley, M. (1989). *The Desktop Fractal Design System*, Cambridge, MA: Academic Press.

33. Çambel, A. B., and Mock, J. E. (1992). "Iterated Function Systems for Geothermal Energy Exploration," *Energy*, **17** (5) 519–522.

CHAPTER **11**

THE DIAGNOSTICS AND CONTROL
OF CHAOS

INTRODUCTION

Chaos per se is neither good nor bad. Under certain circumstances it may lead to the unexpected failure of a bridge or a glitch in the operation of a computer system, or it may endanger a living organism. Chaos may also be beneficial. First, it can serve as a marker, warning us about an impending possibility, thereby giving us an opportunity to bring a situation under control. Second, just as order can deteriorate into chaos, chaos can lead to order. This has been explained by Professor H. Haken,[1] the founder of the Stuttgart School of Synergetics, by Nobelist Ilya Prigogine,[2] and by the urbanologist Jane Jacobs.[3] The concept of free will is related to chaos.

I am referring, of course, to chaos in the technical sense, not to the word that we carelessly toss around in our daily conversations when we complain about disorganization, inefficiency, slovenliness, and most any messy state of affairs. In this volume we are interested in that type of chaos that Professor Katherine Hayles[4] of the University of Iowa has characterized as "orderly disorder." "Orderly" because the systems are deterministic, and "disorderly" because the end result is unpredictable.

We know that whereas chaos is a condition, chaos theory is the amalgam of methods useful to scrutinize nonlinear, dissipative, deterministic problems that have randomness embedded in them. The theory is broad and it is useful whether or not the problem being analyzed turns out to be chaotic. Who has not had his or her temperature taken, only to find out with great relief that there was no fever. That did not make the thermometer a useless instrument!

The outcome of chaotic events is unpredictable in the long term, but chaos theory can provide insights into near-term events. Heretofore, we have learned that a chaotic attractor is associated with the dynamics of such systems. It would be convenient if we could recognize a known attractor, e.g., the attractors of Duffing, Lorenz, Rössler, van der Pol, Shaw, and others. Even if we are unable to recognize a familiar attractor we might be able to simulate it on the computer. We might then be able to arrive at a set of differential equations that serve as a model. From these equations it might be possible to develop the algorithms that separate the chaos signal from the noise signal. Armed with this information one might be able to design a means to suppress the randomness locally. This would result in not having to overdesign, thereby achieving economies without compromising the safety of the structure or the proper functioning of systems. This is not an idle dream—it is a necessity. Our resources are depleting, but our aspirations are increasing. We must learn how to do more with less without sacrificing the quality of life.

It is important to recall that chaos occurs in systems that are sensitive to initial conditions; even a very large system may become chaotic if at some location a minuscule stimulus perturbs the system. This is why meteorologists speak of the *butterfly effect,* the concept that a butterfly batting its wings in Peking stirs the air and thereby causes instabilities in the system to magnify so much that a storm develops weeks later in New York.

If the existence or potential of chaos can be diagnosed, two strategies can be considered: (a) The impending chaotic condition is undesirable and should be moderated; (b) the chaotic condition is of a type that can be exploited to advantage. This is why it is necessary to be able to diagnose whether or not the system is chaotic. This is not simple. There is no instrument that we can stick into the system like a thermometer to ascertain whether or not the system has passed into the chaotic regime. Instead there is a variety of techniques that must be utilized to establish the presence of chaos. Even if we establish the likelihood of a chaotic condition, it does not mean that one must take immediate steps one way or another.

The objective of this chapter is to review how the occurrence of chaos can be diagnosed, and how it might be exploited to exert control over the chaotic condition.

TIME SERIES

Data may be obtained and presented in numerous ways. The lottery data that we saw in Chapter 1 were presented as a scatter plot. Another scatter plot would result if we were to walk into a concert hall and, having obtained the conductor's permission, inquire about the age of each person in the audience. It can be expected that the ages would be statistically independent. In other words, they do not follow any particular order. Chances are that there would be some bias; most likely babies and very old persons would be underrepresented.

In contrast, there are certain data groups that follow a prescribed sequence. If the data are ordered as a sequence in time, we speak of a *time series*. The interpretation of such data plays an important role in determining the existence of chaos. Time series are very common. The graphs of the weekly, monthly, or yearly Dow Jones averages that appear in the business section of the newspaper are typical time series. See Fig. 1 for a typical example.

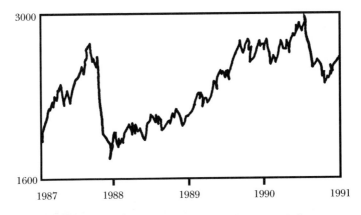

Figure 1. Dow Jones average during five-year period.

Graphs of temperature variations are also time series. Different time series may or may not look alike. Further, it is important to know the background underlying the data. Let us look at two different time series of carbon dioxide. Carbon dioxide in the atmosphere is an important issue because it is a major greenhouse gas, a potential cause for the rise of the atmospheric temperature of the earth. Obviously, this is a concern that evokes fear in all of us. In Fig. 2, kindly provided to me by Thomas A. Boden of the Oak Ridge National Laboratory,[5] may be seen the emissions of carbon dioxide into the atmosphere due to fossil fuel consumption, cement manufacturing, and gas flaring.

It is unlikely that Fig. 2 exhibits chaos, because the sources of the emissions come from specific industrial equipments that are designed for specific operating ranges, and these may be expected to be operated in a controlled manner. Hence there should be no randomness. This assertion is reinforced by the configuration of the curve, which does not appear to be chaotic. Just because CO_2 is emitted into the atmosphere does not mean that it will reside there, because there are many sinks that can absorb the gas.

In contrast, Fig. 3, also from Boden, Sepanski, and Stoss, looks like a better candidate for a chaotic time series. This is the atmospheric CO_2, not the emissions. The accumulation of greenhouse gases in the atmosphere is not well understood, and there are numerous potential causes for nonlinearities and uncertainties to be introduced. Also, fluctuations are evident.

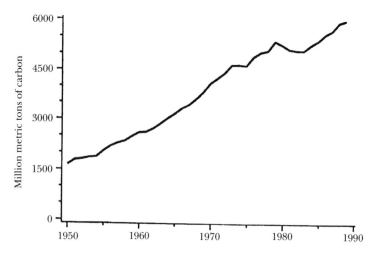

Figure 2. Global CO_2 emissions (source: T. A. Boden, R. J. Sepanski, and F. W. Stoss).

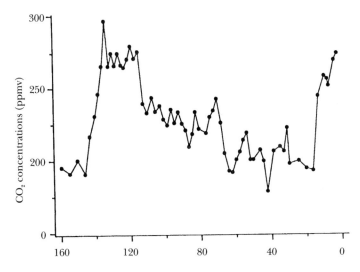

Figure 3. Atmospheric CO_2, derived from the Vostok ice core (source: T. A. Boden, R. J. Sepanski, and F. W. Stoss).

Nevertheless we cannot draw any conclusions until we learn how to analyze time series rigorously. Not every fluctuating time series is chaotic.

Figure 4, kindly provided by Dr. Hiro Kanamori, director of the Seismological Laboratory of the California Institute of Technology, shows two seismograms. The first is for a normal period, while the second shows the traces for the 1991 Sierra Madre earthquake.

The configuration of time series curves depends on the particular problem. It should not be inferred that the absence of fluctuations is always desirable, as in the seismological situation. For example, a constant trace for the Dow Jones average would mean no financial activity, which is quite unlikely to happen. More interesting are physiological time series. The upper portion of Fig. 5 shows the normal rhythm of a healthy individual, while the lower strip is the trace showing abnormal beats indicative of heart disease.

Using clinical data, Dr. Ari Goldberger, of Harvard University and Beth Israel Hospital in Boston, and his associates have shown that contrary to conventional wisdom the heart behaves erratically when young and healthy. Decreased variability and accentuated periodicities seem to be associated with aging and or disease (Goldberger, Rigney, and West[6]). It has also been noticed by Professor Paul Rapp and his associates at the Medical College of

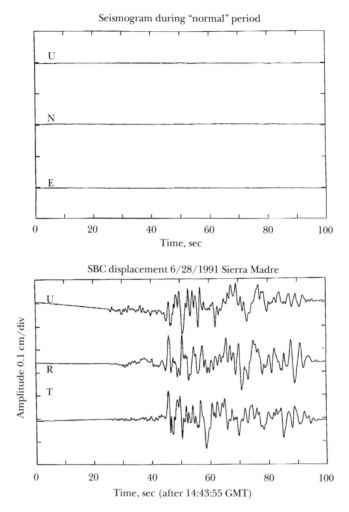

Figure 4. Normal seismogram (upper); Sierra Madre earthquake (lower) (source: H. Kanamori, CalTech Seismological Laboratory; reproduced with permission).

Pennsylvania that there is less variability in neurophysiological disorders (Rapp, Bashore, Zimmerman, Martinerie, Albano, and Mees[7]). Not being a physician I shall refrain from presenting further details, particularly because in life and death matters nobody should be given misconstrued hopes. However, persons who wish to get better acquainted with the

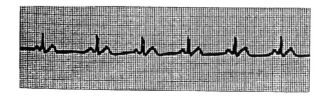

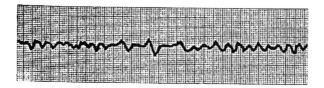

Figure 5. Normal heart rhythm (upper trace) and abnormal rhythm (lower trace).

scientific aspects of chaos in medicine would benefit from the publications by Goldberger *et al.* and Rapp *et al.*, as well as from the books by Glass and Mackey[8] and West.[9]

TIME SERIES ANALYSIS

Analysis of seismic data is of interest to the geophysicists, just as reading an ECG is of interest to the cardiologist, market fluctuations to the stock-broker, and temperature variations to the climatologist. Such examples of time series may be analyzed in several ways: by Fourier analysis, spectral analysis, or power analysis. These are interrelated. If a differentiation is to be made, a Fourier analysis is performed with periodic phenomena, spectral analysis with random phenomena, and power analysis with transient phenomena. However, this pairing-off is not sacrosanct.

There are several reasons for the efficacy of analyzing time series. For example, analysis can be helpful in attempting to predict the future by interpreting past performance. This must be done carefully because in certain situations the form of the data changes. For example, financial data figures may appear higher if there is inflation. Also, predicting the future from the past is not always wise. While there is much to the inscription on the National Archives Building in Washington to the effect that "History Is Prologue," it is also true that with new technologies, and changing population dynamics, the future will not be what it used to be! Time series analysis can also be useful for revealing aspects of the mechanism underlying the configuration of the series and thereby presenting possibilities for control of the causes for undesirable fluctuations. Fatigue in technological devices and structures comes to mind. Another application is to clarify whether the appearance of a disease is isolated or whether it might lead to an epidemic.

A line graph that constitutes a time series is obtained by connecting time increments, e.g., seconds, minutes, hours, days, weeks, years, etc. While the data representing the points are as accurate as the measurements, points along the connecting lines are not necessarily exactly correct, because the events may not be represented within the time increments. In other words, the measurement does not have sufficient resolution and/or a fast enough response time. As a general rule of thumb, one should make the number of points, $\{n\}$, large, and the time to make the measurement, $\{t\}$, as short as possible. Attempting to do so is not without its problems.

In general a time series $\{x_t\}$ can be represented by the following type of equation (Anderson[10]):

$$x_t = f(t) + p_t \quad \text{for } t = 1, 2, \ldots, n \tag{1}$$

Here, $\{f(t)\}$ is the deterministic part and $\{p_t\}$ is the stochastic, or random, component. Recalling our discussion about entropy, $\{f(t)\}$ represents the signal, and $\{p_t\}$ the noise. A major problem is to be able to identify the makeup of the stochastic term. Is the randomness associated with deterministic chaos? Is it due to measurement error? Is it just noise impacting on the system? Regarding the deterministic component, $\{f(t)\}$, one will want to know how it varies with time. If this is periodic we talk of a Fourier series.

Fast Fourier Transforms

Fourier analysis has a long and distinguished tradition in gaining insights into complex problems. Such computations were very formidable, but have become quite reasonable with the development of fast Fourier transforms (FFT) by Cooley and Tukey,[11] and, of course, by the invasion of the ubiquitous computer. It is outside the scope of this volume to present the details of Fourier analysis; the interested reader should consult detailed treatments such as that by Ramirez.[12]

Performing a FFT basically works this way: One starts with the Fourier transform for a smooth function, and then expresses it in discrete form.

The algorithm for the discrete Fourier transform is the FFT. The advantage of FFTs is that the number of computer operations, i.e., the number of multiplications and additions, is reduced materially. Traditionally the number of computer operations goes up as the square of data points, because there are $(N-1)^2$ multiplications and $N(N-1)$ additions (Walker[13]). Even with as few as 1,000 points, this can be time-consuming. However, by resorting to FFTs this large number of operations is reduced to $N \log_2 N$ operations (Baker and Gollub[14]). It is evident that invoking FFTs is very advantageous. One software package dedicated to FFTs has been developed by Falter,[15] and the mathematics software package MathCAD[16] also contains functions to perform FFTs. Those who desire preforming nonlinear forecasting will find much use in the software package by Schaffer and Tidd.[17] Those readers who insist on writing their own programs would get a good start from the FFT program listing by Baker and Gollub. Undoubtedly there are others, but I shall let the respective programs speak for themselves.

Changes in dynamical systems over a period of time can be represented in two different ways: continuous and discrete. My son, who is a communications professional, has a dual-trace oscilloscope hooked up in parallel with the loudspeakers of his hi-fi system. I don't know which he enjoys more, the wafts of beautiful music, or the visual complexity of the traces on the oscilloscope screen. As we saw earlier, even a single note on one musical instrument can have a complex, continuous trace. My hi-fi system is a run-of-the-mill unit bought off the shelf. Instead of an oscilloscope there are two narrow windows through which I can see green and red bars that vary in length as the loudness changes. These are an indication of the electronic power. Over a period of time these bars make up the *power spectrum*. By definition the power spectrum of a field is the square of the Fourier transform of that field. The reason for preferring the power spectrum in many applications is because generally the Fourier transform is complex-valued, i.e., it has imaginary components. By squaring it, one obtains the real function power spectrum. The power spectrum, as well as the Fourier transform, is very useful in analyzing time series data associated with dynamical systems. The analysis of time series may be performed using either frequency spectra or power spectra. The choice becomes important when one tries to tease out all of the information that one can from the available data.

In Chapter 1, sketches by Richard Feynman depicted changes of the flow patterns as the Reynolds number increased. Also, in Eq. 1 of this chapter we expressed time-dependent fluctuations in a flow. Another way of looking at the problem is to study the power spectrum of the flow. In doing so I shall follow the exposition of Professor Leo Kadanoff[18] of the Fermi Institute at the University of Chicago. In one of the seminal papers in the field of chaos theory he defines the power spectrum of the velocity field $\{P(\omega)\}$ by the following equation:

$$[P(\omega)]^{1/2} = \hat{V}(\omega) = \frac{1}{\sqrt{T}} \int_0^T dt\, e^{2\pi i \omega t}\, V(t) \tag{2}$$

In Fig. 6 from Kadanaoff are shown the sequences when the dimensionless Reynolds number, $\{Re = VL\rho/\mu\}$, is increased.

The first column depicts the physical pattern, the second column displays the velocity history, and the third column shows the power spectra. In the first row the Reynolds numbers is very low, and the flow is smooth and laminar. The power spectrum has a spike at a frequency of zero. This configuration prevails even when the Reynolds number is increased slightly, provided the flow is time-independent and smooth downstream. At sufficiently high

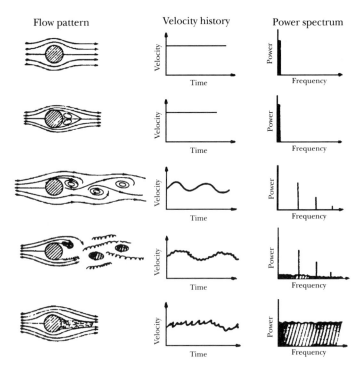

Figure 6. Flow patterns with varying Reynolds number, velocity history, and power spectra. (Source: *Roads to Chaos* by L. Kadanoff. Reproduced with permission of the author.)

Reynolds numbers (third row) the flow changes character and assumes periodicity, while spikes appear at the frequency of oscillation and at the harmonics. As the Reynolds number is increased still further, Kármán vortex-shedding commences, and is probably involved with chaos. In the fourth row, a mixture of spikes and a broad band appear. Finally, when the Reynolds number is so high that the flow becomes completely chaotic, the power spectrum becomes continuous. The reader would be well advised to study Professor Kadanaoff's paper in its entirety for its insightful explanation of power spectra. For an alternate explanation of the different kinds of spectra, the reader is referred to the volume by Bergé, Pomeau, and Vidal.[19]

A major issue involving time series analysis that is not well settled is the number of data points that must be considered. How much of a trace is needed to draw any meaningful conclusions? Can a cardiologist be satisfied with one short rhythm taken during an office visit, or should the patient be provided with a Holter ECG that will record the rhythms over a number of hours? There is no general answer, and each case must be treated on its own merits because different phenomena have different characteristic traces. Within each class of problem one must know the duration and the frequency of extraordinary events. For example, Goldberger, Rigney, and West[20] have shown that for a healthy individual fluctuations of the heart rate appear to be similar when recorded for 3, 30, and 300 minutes. In a cavalier manner one might generalize this to other dynamic problems and suggest that it does not make much difference. I don't recommend this, because how can one be sure that the phenomena have the same scale dimensions over different periods of time? It also depends on the issue being studied. For very high-speed phenomena, there may not be much choice but to be satisfied with little data. Of course, it must be determined that there are sufficient data points to be statistically significant. On the other hand, using an inordinately large data base just because data are available may be counterproductive because the time series would no longer be correlated. Arbitrarily increasing the number of points by slicing the time series more closely just to have a nice-looking data base cannot be recommended either, because it would be meaningless and certainly not very scholarly. It might make things appear more precise, but it does not add accuracy, because of inherent delays in the system. Probably, the best approach is to use two criteria: (a) What type of problem are we dealing with? For example, for studies of market volatility a longer period, such as decades, may be advisable (Peters[21]). On the other hand, for evaluating tool bit performance it may be more like minutes or hours (Basile[22]). (b) The second criterion involves which particular indicator of chaos is being sought. An often-cited reference to determine Kolmogorov entropy is that of Grassberger and Procaccia,[23] who used 12,000 points. Wolf, Swift, Swinney, and Vastano[24] developed a method for determining Lyapunov exponents from a time series. They concluded that for a Hénon attractor the range of data points is 30–100 points; for a Rössler attractor 100–1,000 points; for hyperchaos 1,000–30,000 points; and for a Lorenz attractor a little over 8,000 points. This shows that the required data points depend on the dimensions of the strange attractor being searched. More recently, Sugihara and May[25] have proposed a method using relatively few points, such as 1,000, which is useful in applications such as ecology and epidemiology. The nonlinear forecasting software program of Schaffer and Tidd[26] handles up to 10,000 points. The trend in recent research seems towards lower numbers. This makes sense because one will want to know how early instabilities are likely to arise.

Another fundamental complication that arises with most time series analyses is the matter of noise. The difficulty arises because it is so difficult to separate randomness from noise. The importance of identifying the random components properly cannot be underestimated. While noise is a random process, we need to differentiate between noisy data, such as errors in measurement, and deterministic randomness, which gives rise to chaos. Of course, the matter is exacerbated by the fact that noise is quite common in dissipative systems.

One way of looking at the randomness or the noise is to differentiate between noise internal to the system and noise that is external. It is the former that is likely to lead to chaos. However, it is difficult to differentiate this from the external noise that may be due to errors in measurement or, worse, simple unawareness of what is to be measured. It must be kept in mind that a time series might appear to be chaotic, but upon further scrutiny can turn out to be periodicity plus noise, which is certainly not advisable for forecasting and decision making. Different techniques for constructing predictive models from chaotic time series have been compared by Casdagli.[27]

Time series are not the only way of correlating data. An old hand that is receiving renewed attention is the *log-normal distribution*.

LOG-NORMAL DISTRIBUTIONS AND $1/f$ NOISE

A major cause of complexity is uncertainty. This assertion is made on the premise that the more complex a system, the more detailed its description must be. That means that more variables are involved—and we might not even know all of them. Hence we lack information, and hence the greater the uncertainty. With uncertainty are associated random variables, $\{x\}$, which can assume a number of unpredictable values. Stated differently, a random variable is one that in different experiments or trials assumes different values, $\{x_n\}$. Typical examples are throwing dice or shuffling cards. For random events that have different $\{x_n\}$'s that correspond to various probabilities $\{p_n\}$, one may write $p_1 + p_2 + p_3 + \ldots + p_n = 1$. In contrast to a discrete random variable, a continuous random variable may have any value over a prescribed continuous range. The probability of a particular value of $\{x\}$ is given by the so-called *probability density function* $\{f(x)\}$. Numerous indicators when considered in large numbers fall along a bell-shaped distribution curve. Traditionally, variables including personal income, IQ scores, class grades, and physical height are expected to follow a normal or Gaussian curve. However, contrary to conventional wisdom, this is not always so. For example, class grades exhibit a bell-shaped curve only when the class size is sufficiently large and the grades are studied over a few years. Of course, no two classes will be alike, because the students and the instructors will be different. Income distribution, too, is skewed. It turns out that certain events are better described by the so-called log-normal distribution (Aitchison and Brown[28]).

Data accumulated in a number of real-life situations show that for an event to occur it is necessary that a number of factors coexist. However, even if requisite factors do coexist, the event may not occur for a number of reasons. For example, we may have overlooked an important factor. It is not infrequent in socioeconomics that a course of action that seemed so promising turns sour because certain factors were treated with nonchalance. Discoveries in medicine about previously unknown genetic and/or environmental contributions to different diseases is another such manifestation. In technology, too, serious failure can occur, as in the case of the Tacoma Narrows Bridge, which in 1940 failed four months after completion. This catastrophe has been ascribed to dynamic instability and the use of linear analysis.

The probability, $\{P\}$, that various factors will have some consequence in a certain period of time is the product of the probabilities, $\{p_i\}$, of the individual factors. Of course, we may not know exactly how important each causal factor is. Accordingly, it is not unreasonable to treat the outcome of the event in a probabilistic manner. We can write

$$P = p_1 p_2 p_3 \ldots p_n \quad i = 1, 2, 3, \ldots, n \tag{3}$$

The log-normal formulation is obtained by taking the logarithms, so that

$$\log P = \log p_1 + \log p_2 + \log p_3 + \ldots \log p_n \qquad (4)$$

Montroll and Badger,[29] Montroll and Shlesinger,[30] and West and Shlesinger[31] have presented a number of examples showing that when certain data are presented on log–log graph paper the data appear as a straight line over a certain range. Beyond that range the straight line assumes the shape of a curve according to an inverse power law of the form $\{1/f^{i}\}$ where $\{f\}$ is a frequency and $\{i\}$ is a positive exponent called the *spectral coefficient*.

$1/f^{i}$-type distributions are important otherwise. Persons who still enjoy a good long-playing record probably remember the annoyance they felt when distorted music emanated from their speakers because the player had not been set to correct speed, e.g., 78 rpm instead of 33 rpm. No matter how great the composer, and how inspired the orchestra, the sound was distorted and intolerable. Increasing or decreasing the volume would not improve matters because the composition you were looking forward to listening to has a characteristic time scale. In contrast, there are sounds that regardless of the speed at which they are played back do not change in character if you adjust the volume control. These have been called *scaling 1/f noise* by Mandelbrot.[32] You have probably run across these when your television loses its audio signal and emits an annoying hissing sound.

In 1975 and 1978 two University of California physicists Richard F. Voss and John Clark,[33,34] studied different compositions such as the First Brandenburg Concerto and Scott Joplin rags and found that the $1/f^{i}$ law served as a powerful way of describing music, speech, and noise. They found that compositions having a frequency generated by $1/f$ sources sounded pleasing, while those generated by $1/f^{2}$ sounded too correlated, and those sounds generated from white noise, namely by $1/f^{0}$ sources, sounded too random.

The well-known author of the "Mathematical Games" columns in *Scientific American*, Martin Gardner,[35] noticed the generality of the Voss–Clark work and explained how it related to fractals. He gave the name *Brownian noise* to the case $1/f^{2}$.

In summary, the following nomenclature is used: $1/f^{0}$ is obviously independent of the frequency, and hence independent of its past. It is called *white noise*. $1/f$ is called *pink noise* and is used in acoustics research. As we noted, $1/f^{2}$ noise is called Brownian noise and is very dependent on the frequency. Finally, $1/f^{3}$ is called *black noise*. It should also be noted that the exponents of power laws may be fractions, and this leads to the inquiry of what might be the relationship to fractals.

The relationship between $1/f$ noise and fractals has been further elaborated by Professor Manfred Schroeder[36] in his far-reaching book. He points out that black noise is encountered in natural and artificial catastrophes like floods, bear markets, and electric power outages. He also explains the relationship between black noise and the Hurst coefficient that we learned about earlier. From this he establishes that the spectral exponent $\{i\}$ and the Hurst exponent H are related as follows: $i = 2H + 1$. In other words, we are dealing with fractal noise. Furthermore, we have seen that different time series exist for different $1/f^i$. The fractal dimension can be determined for any such time series. Hence geometric fractal dimensions and temporal fractal dimensions can be related.

We defined a complex system as one being someplace between a completely deterministic and a completely random process. A different description would be a system that fluctuates or has oscillations. The simplest dynamic system is a harmonic oscillator having a single frequency. On the other hand, complex dynamic systems are represented by random time series having numerous oscillations with different frequencies. The spectrum of these frequencies is therefore quite broad. One may then speak of scale invariance if the ratio $\Delta f/f$ of different events is the same. Scale invariance is a characteristic of deterministic fractals. This type of $1/f$ scaling, when joined with the types of scaling that we observed in previous chapters, has very broadly ranging implications. To demonstrate this assertion I wish to note the studies of geophysicist Kenneth Hsü[37] and his musician son, Andrew Hsü, who analyzed compositions of Bach and Mozart and noted fractal geometry.

There are a number of other $1/f$-type distributions that are helpful in evaluating complex data. These include Zipf's law[38] $(-1/n)$, which is revealing in linguistics and social dynamics; Lotka's law[39] $(-1/n^2)$, widely used in problems related to competition; and Pareto's law[40] $(-1/n^v)$ which is used in economics.[41] In all of these $\{n\}$ is the number of occurrences of the case under consideration. For interesting numerical case studies please see the informative paper by West and Salk.[42]

ISING MODEL

In formulating techniques to deal with complexity and chaos we have relied on mathematical models. Here I shall mention a physical model that is proving to be quite helpful in a number of areas where cooperative phenomena are involved. Specifically, I am speaking about the Ising model[43,44] of ferromagnetism that is empirical in character. This describes macroscopic magnetism on a microscopic level. The atoms of magnetic objects constitute minute magnets called *spins*. The macroscopic magnetism is due to the alignment of the spins. When a ferromagnetic object is heated, it loses its magnetism because the spins point in random directions and their magnetic moments cancel out one another. When the ferromagnetic object is cooled, it regains it magnetism because the spins get realigned.

The *Ising phenomenon* is useful in complex systems analysis because it involves directionality and cooperative behavior among the subsystems. Furthermore, the phenomenon is a manifestation of phase transitions. Refreshing applications about the behavior of schools of fish, firefly flashing, and human imitation such as switching soap brands have been discussed by Callen and Shapero.[45] The model has been used successfully in applications as diverse as biology, chemistry, economics and solid state physics. Another example of the application of the Ising model is found in the paper by Pena-Taveras and Çambel,[46] who compared different types of energy investments in the manufacturing sector, with the aid of the Schumpeter clock model, for nonequilibrium, stochastic conditions. Different types of investors were considered, such as the leader and the follower. It was found that had the same problem been pursued with linear models the results would have been off by 56%, because the linear models do not recognize the fluctuations of the variables. Retrospective comparison with historic economic data confirmed the models validity.

CHAOS DIAGNOSTICS

Not all complex systems are chaotic. Here, I shall make an attempt to summarize how one might proceed in looking for the existence of deterministic chaos.

The diagnosis of chaos is not a simple task, because a number of different determinations are necessary. There is no one measurement or calculation that can establish the existence or absence of chaos. For a system to be technically chaotic, certain specific conditions must prevail. These include:

a. The system must be nonlinear and its time series should be irregular.
b. Random components must exist.
c. The behavior of the system must be sensitive to initial conditions.
d. The system should have strange attractors, which generally means that it will have fractal dimensions.
e. In dissipative systems the Kolmogorov entropy should be positive.

f. Perhaps the most terse way of pronouncing a system to be chaotic is to determine that there are positive Lyapunov coefficients.

By definition, *deterministic chaos* means integral presence of randomness, which is a stochastic issue. In stochastic processes the parameters of the system evolve with time, not deterministically, but probabilistically. This leads to uncertainty in making predictions. Accordingly, any model must include probability distributions and various other statistical factors. Until chaos theory came along, modelers resorted to either deterministic models or stochastic models, usually the former to solve physical problems, and the latter to solve social problems. With the advent of chaos theory, the coexistence of the two approaches had to be reconciled.

There are two other matters that must be considered. First, is the system dissipative or conservative? If it is the former, various aspects of entropy may have to be considered because dissipative systems can lead to self-organization. It was said previously that very large systems, such as astrophysical configurations, can be treated quite adequately as conservative systems. Recent developments in cosmology that treat astrophysics in a holistic manner cast doubt on this.

The other issue that is of great importance in practical problems is whether the stochastic component is the randomness inherent in chaos or whether it is noise, or whether it is a combination of both.

In general, the following steps should be followed to determine whether or not the problem is chaotic.

1. Obtain a time series of the events. If this is linear or there are no fluctuations, there is no chaos. Please be aware of the scale to which the data are plotted because it is possible for the data to appear to be linear.

2. If there are patterns, look for random inputs to the system that can appear as chaotic output. We noted earlier that there are a number of alternatives and that the output signal depends on the type of the system and the type of the input signal. As a rule of thumb, chaos may be suspected if the amplitude of the output signal is considerably larger than that of the input signal.

3. Perform a Fourier analysis and obtain the power spectrum. If there is spectral broadening, one may be suspicious that there is chaos or some sort of other randomness. Please see Fig. 6 in this chapter.

4. For a clue to the source of spectral broadening, obtain the *autocorrelation function* $\{R(\tau)\} = \Sigma\, f(t)f(t+\tau)$. What this does is compare a time series with a similar one that has been delayed by

{τ} seconds. If the system is chaotic its correlation function will decay, tending to zero for large values of {τ}. This means that the signal is correlated only with its very recent past. This is why, in the long term, chaotic systems are unpredictable.

5. Because strange attractors are fractal and are associated with chaos, one may conclude that there is chaos for noninteger dimensions. This geometric approach can be extended. Thus, if one places slices through attractors, one obtains Poincaré sections that are manifested by points in the phase plane.

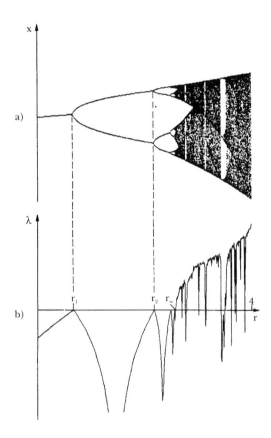

Figure 7. Geometric relation between the bifurcation map and the Lyapunov exponents. (Source: H. G. Schuster. Reprinted with permission of H. G. Schuster and W. Desnizza.)

6. As we noted it is important that the Lyapunov exponent and the Kolmogorov entropy be positive. In Fig. 7, from the instructive book by Professor H. G. Schuster,[47] the logistic map (a) can be seen on top, with the corresponding values of the Lyapunov exponents (λ) below in (b).

The above are signs of the existence of chaos, but they are not sufficient conditions to ensure chaos. For example, in analyzing a set of data points one might encounter a positive Lyapunov exponent. That does not mean that the system is inherently chaotic. It is evident that the diagnosis of chaos is an elaborate process. Under ideal circumstances one would want to satisfy each condition, but this is not always possible, e.g., certain data may not be available. However, by conducting as many of these tests as possible, one can develop insights about the dynamics of the system.

Application

Here I want to present as an example how one might determine whether or not ventricular fibrillation is chaotic. The diagnostic procedure that

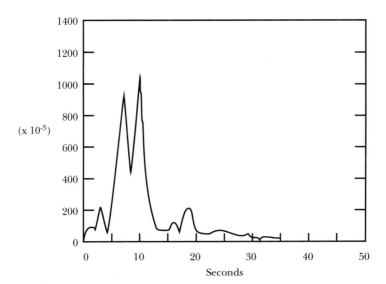

Figure 8. First minute of ventricular fibrillation. (Source: P. Kurzyna. Reproduced with permission.)

follows was worked out by my graduate student, Peter Kurzyna,[48] and is summarized here with his permission. The ventricular fibrillation traces that he used were taken from the reference on electrocardiography by H. J. L. Marriot, M.D.[49]

In addition to checking the various indicators of chaos, three major tests were performed: (a) checking for sensitivity to initial conditions from the Lyapunov exponent of the time series; (b) examining the Fourier spectrum; and (c) determining the fractal properties by means of the Poincaré map and fractal dimensions.

The Fourier spectrums for the first and second minutes were computed and are shown in Figs. 8 and 9, respectively. In both figures a dominant peak is noticeable, indicating possible chaotic motion.

Data were analyzed to identify the noise in the system using J. Falter's FFT program.[50] This is shown in Fig. 10.

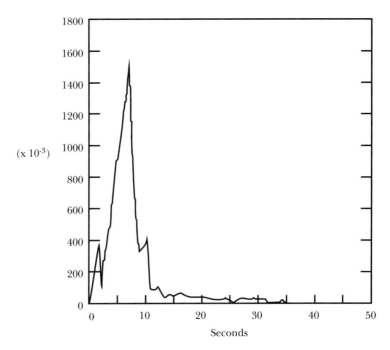

Figure 9. Second minute of ventricular fibrillation. (Source: P. Kurzyna. Reproduced with permission.)

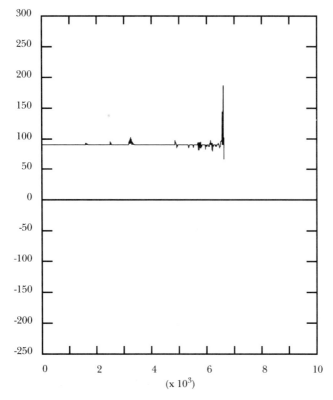

Figure 10. The FFT for ventricular fibrillation. (Source: P. Kurzyna. Reproduced with permission.)

In this figure the horizontal axis represents the first minute of fibrillation, and the vertical axis the second minute.

The Lyapunov exponent was calculated using the program by Wolf, Swift, Swinney and Vastano[51] cited earlier. Among the computational results for the Lyapunov exponent, two were found to be positive, suggesting chaos. The Lyapunov dimension was calculated and found to be 1.15, which suggests the existence of a strange attractor.

The fractal characteristics were studied by plotting the Poincaré map and comparing with other cases where chaos had been noticed. This is shown in Fig. 11.

It is conceivable that the chaos that was diagnosed in this example was fortuitous. D. T. Kaplan and R. J. Cohen[52] studied surface ECGs from dogs

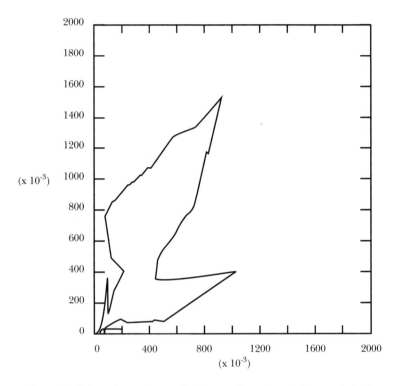

Figure 11. Poincaré map (source: P. Kurzyna. Reproduced with permission).

in fibrillation, but their analysis did not identify an attractor. They concluded that ventricular fibrillation was not chaotic in the technical sense. At its November 1989 annual meeting, the American Heart Association held a seminar to resolve some of the issues involved, but as can be expected there are conflicts, such as contradictions in terminology, paucity of data, poor experimental design, and a host of others matters. It is salutary that the debate fuels excitement in research, but it must be realized that relying on preliminary research data for clinical use is premature. It is quite evident that much work remains to be done before chaos theory can be applied clinically with confidence. This view is corroborated in a concise paper[53] by Dr. Michiel J. Janse, of the Department of Clinical and Experimental Cardiology at the University of Amsterdam, who reviewed the state of the art and made the following statement:

... our insight into arrhythmogenesis will be advanced by the application of chaos theory only if the link between "modern" chaos theory and "old fashioned" electrophysiology is established. Here, [the collaboration of] a mathematician who can model iterative nonlinear systems and who understands the language of the cardiac electrophysiologist and of [a] practicing cardiologist may be of help.

This statement clearly advocates interdisciplinary cooperation in applying chaos theory.

CHAOS CONTROL

As the field of nonlinear dynamics advances, there is an increasing awareness that chaos is actually quite common. Whereas at one time we used to shy away from nonlinear, nonequilibrium problems, we have come to realize that we can deal with them intelligently, albeit not conclusively as yet. This makes it possible to describe the behavior of chaotic systems, as long as we do not predict a guaranteed outcome. Recently, a very major step forward was taken. Specifically, Edward Ott, Celso Grebogi, and James A. Yorke,[54] of the University of Maryland, discovered that chaos can be controlled without having to alter the system completely. What they discovered was to lock the random characteristic of the chaotic system into periodic behavior. Understandably this capability is very attractive because rather than undertaking major changes in the system so that its objectives are compromised, one can tweak the system at appropriate locations and inhibit chaos from occurring. Within less than a year of the theoretical paper, W. I. Ditto, S. N. Rauseo, and M. L. Spano, [55] of the Naval Surface Warfare Center, succeeded in controlling chaos experimentally in the laboratory. They worked with magnetoelastic materials that exhibit nonlinear reactions to applied forces. A ribbon made of such material may stand erect, bend over, or undulate, depending on the strength of the applied magnetic field. They showed that by making adjustments in accordance with the theory, the motion of the ribbon could be made uniform. These developments are providing an impetus for a wide variety of research that could have far-reaching applications. Already Shinbrot, Ott, Grebogi, and Yorke[56] have shown that in a short time one can direct a system to a desired accessible state. A host of experiments and applications are discussed in the volume by M. F. Shlesinger, W. L. Ditto, L. Pecora, and S. Vohra.[57]

Here I can best summarize the existing state of the art by quoting directly from the aforementioned pioneering paper by Ott, Grebogi, and Yorke.

> . . . a chaotic attractor typically has embedded within it an infinite number of unstable orbits. . . . We first determine some of the unstable low-period periodic orbits that are embedded in the chaotic attractor. We then examine these orbits and choose one which yields improved system performance. Finally, we tailor our small time-dependent parameter perturbations so as to stabilize this already existing orbit.

The interesting point is that in a chaotic system numerous unstable orbits exist, and one can stabilize the one most conducive to achieving the desired system stability by imposing only a small stimulus. Nonchaotic systems are not sensitive to low-level stimuli. As a rule, when a nonchaotic system is to be controlled or regulated, the controller is monofunctional, and special arrangements must be made for each function. In contrast, multiple-function systems that are chaotic may be endowed with flexibility. In other words, we take advantage of the existence of chaos to control the system in a manner to our liking, and this control setting is not fixed; there is flexibility. For example, a heart pacer supplies electrical impulses to ensure that the heartbeat is regular. However, a person's heartbeat varies under certain circumstances, such as physical exertion or psychological anxiety. As mentioned earlier, it has been shown by Dr. Ary Goldberger[58] that a very regular heartbeat is not necessarily desirable. The recent theoretical and experimental research in chaos control suggests using chaos as a marker. A pacemaker based on chaos control could conceivably send self-adjusting electrical pulses enabling the heart to beat at intervals dictated by the temporal physiological needs. At this point one can only speculate about the potential uses of chaos control.

Other potential applications of chaos control may be useful in electric power generation and transmission. Under consideration by the Electric Power Research Institute are voltage collapses, electromechanical oscillations, and unpredictable behavior in electric grids.[59]

Chaos control is being studied in chemically reacting systems (Peng, Petrov, and Showalter[60]). Such research could lead to enhanced productivity in chemical processing and could be useful in improving chemical combustion.

There is no question that the implications are far-ranging. Indeed Ott, Grebogi, and Yorke[61] have stated:

> Such multipurpose flexibility is essential to higher life forms, and we therefore speculate that chaos may be a necessary ingredient in their regulation by the brain.

START YOUR OWN CHAOS LABORATORY

It is said that Louis de Broglie, who received the Nobel Prize in physics, liked to work on his estate rather than spending his time at the university. You, too, may wish to do so. More seriously, the best way to master any subject is through hands-on experience. You might also wish to develop a laboratory in your own educational institution for teaching and research, or for research in your place of employment in industry or government. Fortunately, it is not difficult to put together a modest laboratory that can be productive, without incurring outrages costs.

To be sure, having a major laboratory with its supercomputers, optics equipment, and expert support staff is to be envied. On the other hand, developing a small chaos laboratory has its advantages. First, seeing and touching the actual thing gives one confidence in being discerning. Second, in a small laboratory one can do most of the work oneself and thus become acquainted with all facets. Also, a small facility is much cheaper to establish, and thus one does not have to depend on large financial support. Last but not least, useful and revealing results can be obtained.

Without endorsing any particular products, let me share with you with what I started and to which I just keeping adding.

Essentially you need the following: a computer, software (or the ability to write software), some demonstration devices, and some videotapes.

Concerning the computer, either an IBM or compatible PC, or a Macintosh, will do just fine. I have an AT clone. A faster PC would certainly be preferable. In any case, a math coprocessor is desirable because you will be doing a lot of iterative number crunching as well as graphics. Finally, I recommend a VGA card and monitor.

If you plan to produce hard copies of your work, I recommend a laser printer. Please be aware that printing graphic images may require some preparation because the program and the graphics card may not match. Thus the shift–print-screen command will not always work. Depending on the software, I use the grab command in WordPerfect 5.1[62] or the Raindrop[63] print screen utility.

If printing does not appeal to you, you can photograph images directly off the monitor screen. This has two advantages. First, you can get color images if you do not have a color printer. Second, you can generate slides directly. However, photographing a computer monitor screen is not as simple as pointing your auto-everything camera at an object and clicking the shutter release. The monitor, the image, your camera, and the film all must be harmonized, so be prepared to fool around with different films and different camera settings until you develop the right touch. It is a good idea to record your settings and match them to the print or slide for archival use. A useful guide when you are starting is a pamphlet that you can obtain from the Eastman Kodak Company.[64]

Then you need software. You can develop your own, obtain some excellent commercial programs, or get them from electronic bulletin boards either free or for a nominal charge. Your minimum inventory of software should include one for performing FFT analysis, one for bifurcations and return maps, and one for fractals. I mentioned a number of these. You can also get site licenses for instructional use.

In the line of physical apparatus you can build your own chaotic toys or buy some in novelty stores. One very instructive apparatus that you can make easily and cheaply is a magnetic chaos pendulum, shown in Fig. 12. You need a handle about 12 in. long; this can be a dowel. At one end, screw in an eye screw. To this attach a "lift wire" that is used to lift the plunger ball in your toilet tank and that you can purchase at your friendly hardware store. While you are there, get half a dozen or so round permanent magnets. Glue one of these to the end of the lift wire with very adhesive glue. Then on a piece of plastic, draw a dartboard, like one with a target, and attach the magnets you purchased.

Place this on an overhead projector, hold the magnetic pendulum about 3/4 in. above the target, and your audience will observe all sorts of chaotic behavior. It is instructive to simulate this physical experiment on a computer, which you can do with the software program entitled *James Gleick's CHAOS: The Software.*[65]

A number of mechanical and electrical chaotic toys are discussed by Professor Francis Moon[66] of Cornell University. Of course, you could also revive van der Pol's original circuit, which I mentioned earlier in connection with attractors. Some toys that I collected at airport gift shops are shown in Fig. 13. Unfortunately, I cannot give their source or proper credit because these came completed, unmarked, and unidentified.

We are surrounded by fractal objects, and it is interesting to identify the fractal dimension of interesting objects. Dividers are convenient because the scale can be changed. Another simple approach is to copy xerographically the fine grid from graph paper on a sheet of transparency and happily go along doing your box counting.

Useful fractal experiments such as viscous fingering in porous media and two-phase flow can be performed with Hele–Shaw apparatus (Walker[67]). A crude Hele–Shaw apparatus can be constructed by placing institutional

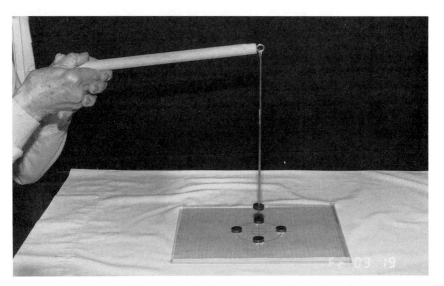

Figure 12. Homemade chaotic pendulum and target board.

Figure 13. Two chaotic toys.

219

paper towels (dry or wet) between two sheets of plastic and injecting fluids with different colors and viscosities.

Finally, there are occasions when people get passively interested in what you are doing. For such occasions I recommend certain videotapes. If you cannot afford to purchase them, you might persuade your library to obtain them. The ones my students and I found useful are the videotape on chaos that was broadcast by the Public Broadcasting System during a NOVA program; a stimulating conversation with Edward Lorenz and Benoit Mandelbrot;[68] an excellent tutorial tape by Robert Devaney;[69] and the mesmerizing videotape produced by Art Matrix[70] in cooperation with the Cornell University supercomputer facility.

So you, too, can be a practicing chaologist. Remember, *we are complexity.* Hence I close by quoting J. D. Farmer, one of the founders of chaos theory:

> Old age might be defined as the onset of limit cycle behavior. May your chaos be always of high dimension.

NOTES AND REFERENCES

1. Haken, H. (1983). *Synergetics,* Berlin: Springer-Verlag.
2. Prigogine, I., and Stengers, I. (1984). *Order out of Chaos,* Toronto: Bantam Books.
3. Jacobs, J. (1969). *The Economy of Cities,* New York: Random House.
4. Hayles, N. K. (1990). *Chaos Bound,* Ithaca, NY: Cornell Univ. Press.
5. Boden, T. A., Sepanski, R. J., and Stoss, F.W., eds. (1991). *Trends '91: A Compendium of Data on Global Change,* ORNL/CDIAC-46, Oak Ridge, TN: Carbon Dioxide Information Center, Oak Ridge National Laboratory.
6. Goldberger, A. L., Rigney, D. R., and West, B. J. (1990). "Chaos and Fractals in Human Physiology," *Scientific American,* **262**(2), 43–49.
7. Rapp, P. E., Bashore, T. R., Zimmerman, I. D., Martinerie, J. M., Albano, A. M., and Mees, A. I. (1990). "Dynamical Characterization of Brain Electrical Activity," in *The Ubiquity of Chaos,* (pp. 10–22), S. Krasner, ed., Washington, D.C.: AAAS.
8. Glass L., and Mackey, M. C. (1988). *From Clocks to Chaos,* Princeton, NJ: Princeton Univ. Press.
9. West, B. J. (1990). *Fractal Physiology and Chaos in Medicine,* Singapore: World Scientific.
10. Anderson, T. W. (1971). *The Statistical Analysis of Time Series,* New York: John Wiley & Sons.
11. Cooley, J. W., and Tukey, J. W. (1965). "An Algorithm for Machine Calculation of Complex Fourier Series," *Mathematical Computation,* **19**, 297.
12. Ramirez, R. W. (1985). *The FFT: Fundamentals and Concepts,* Englewood Cliffs, NJ: Prentice-Hall.
13. Walker, J. S. (1988). *Fourier Analysis,* Oxford: Oxford Univ. Press.
14. Baker, G. L., and Gollub, J. P. (1990). *Chaotic Dynamics,* Cambridge: Cambridge Univ. Press.
15. Falter, J. J. (1988). *Fast Fourier Transform Program,* Adelphi, MD: Harry Diamond Laboratories.
16. MathSoft, Inc., Cambridge, MA 02139.

17. Schaffer, W. M., and Tidd, C. W. (1991). *NLF: Nonlinear Forecasting for Dynamical Systems,* Tucson, AZ: Dynamical Systems Corporation.

18. Kadanaoff, L. P. (1983). "Roads to Chaos," *Physics Today,* December, 46–53.

19. Bergé, P., Pomeau, Y., and Vidal, C. (1984). *Order Within Chaos,* New York: John Wiley & Sons.

20. Goldberger, A. L., Rigney, D. R., and West, B. J. (1990). *op. cit.*

21. Peters, E. E. (1991). *Chaos and Order in the Capital Markets,* New York: John Wiley & Sons.

22. Basile, S. (1992). Personal communication.

23. Grassberger, P., and Procaccia, I. (1983). "Characterization of Strange Attractors," *Physical Review Letters,* **50**(5), 346–349.

24. Wolf, A., Swift, J. B., Swinney, H. L., and Vastano, J. A. (1985). "Determining Lyapunov Exponents from a Time Series," *Physica,* **16D**, 285–317.

25. Sugihara, G., and May, R. M. (1990). "Nonlinear Forecasting as a Way of Distinguishing Chaos from Measurement Error in Time Series," *Nature,* **344**, 19 April, 734–741.

26. Schaffer, W. M., and Tidd, C. W. (1991). *op. cit.*

27. Casdagli, M. (1989). "Nonlinear Prediction of Chaotic Time Series," *Physica D,* **35**, 335–356.

28. Aitchison, J., and Brown, J. A. C. (1957). *The Lognormal Distribution,* Cambridge: Cambridge Univ. Press.

29. Montroll, E. W., and Badger, W. W. (1974). *Introduction to Quantitative Aspects of Social Phenomena,* New York: Gordon and Breach Science Publishers.

30. Montroll, E. W., and Shlesinger, M. F. (1982). "On $1/f$ Noise and Other Distributions with Long Tails," *Proc. Nat. Acad. Sci.,* **79**, 3380–3383.

31. West, B. J., and Shlesinger, M. F. (1990). "The Noise in Natural Phenomena," *American Scientist,* **78**, 40–45.

32. Mandelbrot, B. B. (1983). *The Fractal Geometry of Nature,* San Francisco, CA: W. H. Freeman and Co.

33. Voss, R. F., and Clark, J. (1975). "$1/f$ Noise in Music and Speech," *Nature,* **258**, 317–318.

34. Voss, R. F., and Clark, J. (1978). "$1/f$ Noise in Music: Music from $1/f$ Noise," *Journal of the Acoustical Society of America,* **63**(1), 258–263.

35. Gardner, M. (1978). "White and Brown Music, Fractal Curves and One-over-f Fluctuations," *Scientific American,* April, 16–32. Also see: Gardner, M. (1992). *Fractal Music, Hypercards and More...,* New York: W. H. Freeman and Co.

36. Schroeder, M. (1991). *Fractals, Chaos, Power Laws,* New York: W. H. Freeman and Co.

37. Hsü, K., and Hsü, A. (1991). "Self-Similarity of the '$1/f$ Noise' Called Music," *Proc. Nat. Acad. Sci.,* **88**, 3507–3509.

38. Zipf, G. K. (1949). *Human Behavior and the Principle of Least Effort,* Cambridge, MA: Addison-Wesley Press, Inc.

39. Lotka, A.J. (1926), "The Frequency Distribution of Scientific Productivity," *Journal Washington Acad. Sci.,* **16**, 317–323.

40. Pareto, V. (1897). *Cours d'Economie Politique,* Lausanne: This original work may be difficult to locate. A helpful description is provided in Montroll, E. W. and Badger, W. W. (1974). *Introduction to Quantitative Aspects of Social Phenomena,* New York: Gordon and Breach Science Publishers.

41. In each case, $\{n\}$ represents characteristic number of items appropriate to the problem.

42. West, B. J., and Salk, J. (1987). "Complexity, Organization and Uncertainty," *European Journal of Operational Research,* **3**, 117–128.

43. Brush, S. G. (1967). "History of the Lenz–Ising Model," *Reviews of Modern Physics,* **39**, 883–893.

44. McCoy, B. M., and Wu, T. T. (1973). *The Two-Dimensional Ising Model,* Cambridge, MA: Harvard Univ. Press.

45. Callen, E., and Shapero, D. (1974). "A Theory of Social Imitation," *Physics Today*, July, 23–28.
46. Pena-Taveras, M. S., and Çambel, A. B. (1989). "Nonlinear, Stochastic Model for Energy Investment in Manufacturing," *Energy—The International Journal*, **14**(7), 421–433.
47. Schuster, H. G. (1988). *Deterministic Chaos*. Weinheim, Germany: VCH, Verlagsgesellscht.
48. Kurzyna, P. (1991). "Another View on the Question: Is Ventricular Fibrillation 'Chaos'?" Term Paper. Special Topics in Chaos, Washington: George Washington University.
49. Marriot, H. J. L. (1975). *Practical Electrocardiography*, Baltimore: Waverly Press Co.
50. Falter, J. J. (1988). *op. cit.*
51. Wolf, A., Swift, J. B., Swinney, H. L., and Vastano, J. A. (1985). *op. cit.*
52. Kaplan, D. T., and Cohen, R. J. (1990). "Is Fibrillation Chaos?" *Circulation Research*, **67**(4), 886–892.
53. I am indebted to Fritz H. Andersen, M.D., for providing me with this paper.
54. Ott, E., Grebogi, C., and Yorke, J. A. (1990). "Controlling Chaos," *Physical Review Letters*, **64**(11), 1196–1199.
55. Ditto, W.I., Rauseo, S. N., and Spano, M .L. (1990). "Experimental Control of Chaos," *Physical Review Letters*, **65**(26), 3211–3213.
56. Shinbrot, T., Ott, E., Grebogi, C., and Yorke, J. A. (1990). "Using Chaos to Direct Trajectories to Targets," *Physical Review Letters*, **65**(26), 3215–3218.
57. Vohra, S., Spano, M., Shlesinger, M., Pecora, L., and Ditto, W., eds. (1992). *Proc. 1st Experimental Chaos Conf.* Singapore: World Scientific.
58. Ary Goldberger, M. D., has been a prolific contributor to the subject of cardiology and chaos. Here I cite one of his publications that includes a rich bibliography on the subject. Goldberger, A., and Rigney, D. R. (1990). "Sudden Death Is Not Chaos," *The Ubiquity of Chaos*, S. Kasner, ed. (pp. 23–34). Washington: American Association for the Advancement of Science.
59. Stambler, I. (1991). "Chaos Creates a Stir in Energy-Related R&D," *R&D Magazine*, December, p. 16.
60. Peng, B., Petrov, V., and Showalter, K. (1991). "Controlling Chemical Chaos," *Journal of Physical Chemistry*, **95**, 1957–1959.
61. Ott, E., Grebogi, C., and Yorke, J. A. (1990). *op. cit.*
62. WordPerfect 5.1 (1989). Orem, Utah: WordPerfect Corporation.
63. Raindrop Enhanced Print Screen Utility (1990). Springfield, VA: Eclectic Systems.
64. Eastman Kodak Company (1986). *Photographing Television and Computer Screen Images*, Kodak Publication No. **AC-10**. Rochester, NY: Photographic Products Group.
65. James Gleick's CHAOS: The Software (1990). Sausalito, CA: Autodesk, Inc.
66. Moon, F. (1987). *Chaotic Vibrations* (pp. 279–286). New York: John Wiley & Sons.
67. Walker, J. (1987). "Fluid Interfaces Including Fractal Flows Can Be Studied in a Hele-Shaw Cell," *Scientific American*, **257**(5), 134–138.
68. Peitgen, H.-O., Jürgens, H., Saupe, D., and Zahlen, C. (1990). *Fractals—An Animated Discussion with Edward Lorenz and Benoit B. Mandelbrot*, (1990). New York: W. H. Freeman.
69. Devaney, R. L. (1990). *Transition to Chaos—The Orbit Diagram and the Mandelbrot Set*, New York: Science Television.
70. Art Matrix. *Nothing but Zooms*, Ithaca, NY: Cornell Univ. Press.

CHAPTER **12**

DISCUSSION TOPICS

The following discussion topics are offered to enhance the understanding of the subjects covered in this volume. I have intentionally called this section "Discussion Topics" because the topics go beyond the usual limitations of homework problems. In many cases there is no single correct answer. This is due to the very nature of the subject. Further, the particular interpretation of the problem statement is quite important. In the classroom I have found that there may be a number of approaches, all acceptable although some are more powerful than others.

Some of the topics can be pursued qualitatively, others quantitatively, and some in both ways. The choice depends on the person's background and tastes. The discussion topics are not limited to any discipline, so any interested person is invited to participate in the discussions.

I have sorted out the topics cited by the content of the different chapters. However, certain subjects can be dealt with in different ways, and the peripatetic reader should not deny himself or herself the pleasure of tackling a subject previously treated when another applicable approach makes its appearance in a later chapter. Indeed, this can add to the excitement when different conclusions come about. One should then ask: "Why did that happen?" and try to get to the root cause. Frequently, this will result in novel insights. Depending on the background of the reader, some of the discussion topics may require gathering information from sources outside this volume. As Thomas Carlyle remarked, "A true university is a good library." While some of the questions may be answered cryptically, others might be the subject of erudite theses, doctoral dissertations, or even research proposals.

Chapter 1—Living with Complexity

1) Differentiate among simple, complicated, and complex.
2) Compare a single cell, a solitary ant, and a lonely human in regard to complexity.
3) Is a rock more complex than a pebble, and is that in turn more complex than a kernel of sand?
4) Is a bacterium less complex than a virus because frequently it can be treated with an antibiotic? Does the statement make sense?
5) Must a system entail randomness to be considered complex?
6) Which is more complex: a bottle of perfume or a robot that sprays paint on car bodies?
7) Which reaction is more complex: a methane oxygen reaction or a deuterium–deuterium reaction?
8) Which is more complex: 5 cc of blood of a healthy individual or 5 cc of mayonnaise kept in a refrigerator? Reinterpret your statement if the mayonnaise has been left out in the sun during a picnic.
9) Must a system be nonlinear and dynamic to be complex?
10) Cite as many actual linear phenomena that one encounters during a typical day. Write and solve the defining equation.
11) Cite all the nonlinear phenomena that you normally encounter during one week. Try to write the associated equation(s) representing the events. Try to present the solution(s).
12) What does Jacques Monod mean when he asserts that chance is not further reducible? (See *Chance and Necessity*, New York: Vintage Books, 1972.)
13) Differentiate among determinism, indeterminism, uncertainty, randomness, and noise.
14) Differentiate between deterministic chaos and noise.
15) Compare the limitations of reductionism with those of holism.

Chapter 2—Meta-Quantification of Complexity

1) Discuss the validity of Kelvin's statement cited in this chapter about the need for measurement.
2) Compare "truth" and "fact."
3) Compare "accuracy" with "precision."
4) In this chapter several hierarchies of complexity were presented. Does it follow that succeeding hierarchical levels are more complex?
5) It was suggested that additional categories appear as time goes on. Contemplate about future levels of complexity that are apt to appear.
6) Compare the Turing machine and the Carnot engine as mental constructs that serve to set limiting conditions.
7) Differentiate between Gödel's undecidability theorem and Heisenberg's uncertainty principle.
8) Does it follow that Chinese is a more complex language than English because it has thousands of characters, whereas the English alphabet has 26 letters?
9) Compare the Morse code with the Braille alphabet as far as complexity is concerned.
10) Is a 386 computer chip running at 20 MHz less complex than a 386 chip running at 33 MHz?
11) Is a plain sheet of paper less complex than when it is folded into four sections?
12) Is a cooked meal more complex than its raw ingredients? What happens after the meal has been eaten and digested?

Chapter 3—The Anatomy of Systems and Structures

1) Cite as many actual isolated systems as you can.
2) Cite as many open systems as you can.
3) Cite as many useful conservative systems as you can.
4) Cite as many useful dissipative systems as you can.
5) Cite as many self-organizing systems as you can.
6) Differentiate among phase space, state space, phase plane, state plane, parameter space, and parameter plane.
7) List the following in order of decreasing complexity: a bar of gold, a piece of iron pipe, a single cell, a solitary ant, and a beehive.
8) The original Bénard experiments were performed in small laboratory apparatus with characteristic dimensions of centimeters.

To what extent can one apply these lessons to meteorological phenomena where the dimensions are much greater?

Chapter 4—Attractors

1) Cite as many examples of limit cycles as you can.
2) Cite as many examples of strange attractors as you can.
3) Suppose that you are the owner of a fishing pond where you allow people to fish. Obviously you will have to replenish the fish supply from time to time. Analyze this problem in terms of the limit cycle concept regarding the most economic frequency and amount of replacement.
4) Write the computer programs for the van der Pol equation and the Duffing equation.
5) Perform the layered fudge (Arnold's cat) experiment using ready-made dough and food coloring that you can readily obtain at your supermarket. Feel free to make any image of your choosing. Should you expect to reconstruct the original image?
6) How might the philosophers Charles Peirce, Karl Popper, and Ludwig Wittgenstein have contributed to our understanding of the subjects discussed in Chapter 4?
7) What, if any, attractors do static systems have?
8) In what ways are fixed-point attractors, limit cycles, and tori different from strange attractors? What similarities are there among these four basic types of attractors?
9) Sketch a basin of attraction.
10) Sketch a torus, and explain the types.
11) What difference, if any, is there between a strange and a chaotic attractor?
12) Sketch the figures a three-dimensional pendulum can be expected to trace out.
13) Combine the van der Pol and the Duffing equations and seek a solution.
14) Why is the Hénon map so interesting? What applications does it have? Does it apply to conservative or dissipative systems? Please explain.
15) Sketch the time series for the Rössler equations. What are potential applications of the Rössler attractor? Does it apply to conservative or dissipative systems? Please explain.
16) Is the system of equations that Lorentz wrote applicable to both conservative and dissipative systems? Please explain.

17) Compare the Poincaré recurrence in electrical and in mechanical systems.

18) What is the relationship between the Lyapunov exponent and the Kolmogorov entropy?

Chapter 5—Rapid Growth

1) Identify those classes of problems wherein Malthus's theory is valid, and those wherein it is not. Explain the underlying reasons.

2) Derive the "Rule of 70," which states that the doubling time of an exponentially growing item is approximately equal to 70/annual percentage.

3) Compare growth phenomena by calculating and drawing the following time series curves: linear, exponential, logistic, and Fibonacci.

4) Was there anything fundamentally wrong with Malthus's reasoning?

5) Compare the viewpoints of Darwin, Lamarck, and Malthus.

6) Can growth phenomena follow an exponential curve without limit?

7) The slope of many exponential curves increases as we move along the curve and, depending on the time, the curve can appear to have become vertical. Under what circumstances can it tip over backwards and assume a negative slope? As an illustrative experiment, plot the costs of a large construction project as the abscissa, and the benefits as the ordinate. Both the costs and the benefits are functions of time, i.e., $B = f(t)$ and $C = f(t)$. Under what circumstances can the curve start out with a negative slope that gradually flattens out, then turns around and assumes an increasing positive slope, which passes through an inflection point and then tips over into a negative slope and continues along a gradually decreasing positive slope?

Chapter 6—The Logistic Curve

1) To what extent do dynamic systems follow random, exponential, and S-shaped patterns?

2) You are presented with an exponential graph that is still in its early period. Determine when this curve will become a sigmoid curve.

3) Do all logistic curves start with an initial exponential portion?

4) Solve Eq. 10 for the following cases: (a) $N_1 = N_2 = N_3 = 0$; (b) $N_1 = N_2 = N_3$. (b1) they are positive, but less than unity; (b2) they are positive, but greater than unity; (c) $N_1 > N_2 > N_3$,

and they are all positive. Try different mixes of less than and greater than unity.

5) How does Eq. 10 apply to mergers and hostile takeovers among companies? Write scenarios that interest you, and analyze each, using the concepts of the logistic curve.

6) Under what conditions might Eqs. 10 be in stable equilibrium?

7) Describe how Eq. 8 can be applied to describe the progress of epidemics.

8) Certain diseases can be caused by more than one factor. Also, several medications may be helpful. Using the various types of logistic equation models, e.g., predator–prey, combat, and competition, attempt to describe the sequence of events.

9) How can you tell whether or not a surge of an infectious disease will lead to an epidemic?

10) Consider the very complex problem of crime prevention. Would increasing the number of law enforcement officers eradicate crime, or would there be a possibility of arriving at a limit cycle situation? Whom does the theory favor: the law enforcers, or the criminals?

11) Can the cicada invasion that occurs in cycles of from 1 to 17 years be studied by the equations discussed in this chapter?

12) What are the limitations of the continuous logistic equation?

13) An infectious disease makes its appearance. Without consideration of cost, what would be more effective: vaccination, quarantine, or waiting it out?

14) You are the holder of the patents, all manufacturing facilities, and have access to the distribution sources of a technological innovation. Should you or should you not make the investment to go ahead, realizing that (a) your device is an improvement over existing technologies? (b) your device is entirely novel and there are no others in its class on the market? Of course, the novelty factor makes it difficult to evaluate the potential market.

15) Use data from lynx–hare populations fluctuations and plot these on the predator–prey plane. In doing so use data over the entire range, as well as for subranges. In each case connect the points. Do you see orbits? Describe and explain your findings.

16) An island has a substantially fixed population except for tourists that come and go. One season a plane brings 10 passengers carrying a fatal infectious disease to which these 10 persons themselves are immune. Any contact between a local person and one

of the passengers causes death to the former within three days.
Set up a mathematical model patterned after the continuous
Lotka–Volterra equations, and discuss the consequences.

Chapter 7—The Discrete Logistic Equation

1) Discuss the pros and cons of the continuous and the discrete
logistic equations.
2) You are presented with a time series. Under what circumstances
would you choose the continuous logistic curve? When would
you choose the discrete logistic equation?
3) Derive the discrete logistic equation from the continuous logistic
equation.
4–9) Repeat discussion topics 8, 9, 10, 11, 13, and 14 of Chapter 6
using the discrete logistic equation.
10) Consider a time series that for some reason is completely deterministic and absolutely devoid of randomness. Would there be a
bifurcation map?
11) Consider a completely random process over time. Would there
be a bifurcation map?
12) What are the limitations of the discrete logistic equation?
13) Perform iterative computations in order to plot the equation
$X_{n+1} = 4X_n(1 - X_n)$, where $0 < X_n < 1$. Perform computations for
$X_n = 0.00, 0.10, 0.20, 0.30, 0.40, 0.50, 0.60, 0.70, 0.80, 0.90$, and
1.00. Then repeat the calculations for the initial $X_n = 0.01, 0.11$,
$0.22, 0.33, 0.44, 0.55, 0.66, 0.77, 0.88, 0.99$, and 1.01. Compare
the two sets by superimposing the different plots. You may have
to do this in groups because of the graphical confusion that
can arise.
14) In the aforementioned problem, what is the influence of the
initial condition?
15) Do you discern chaotic behavior in the previous two problems?
16) Repeat the basic idea of Problems 13, 14, and 15, for the logistic
equation where $0 < X_n < 4$.
17) Using computations and graphics, determine whether or not chaotic
conditions will arise in the difference equation $X_{n+1} = X_a(1 - X_n)$
18) Develop your own difference equation that is unlikely to exhibit chaos.
19) Let us assume that we are asked to make projections in population dynamics for the next two generations. Compare the results
that the discrete logistic equation and the Fibonacci series would
yield. Repeat the problem for the next 10 generations.

20) Think of a pet interest of yours that involves a time series. For example, it could be the speed of airplanes, the speed of champion skiers, the javelin or discus-throwing records at the Olympic games, the growth of the gross national product, the growth of R&D funding. Project what one might expect a quarter of a century later. How reliable do you expect the results to be?

21) Repeat discussion topic 16 from the previous chapter, using the discrete logistic equation.

Chapter 8—The Different Personalities of Entropy

1) Which has a higher entropy: a single cell, a single ant, or a single individual? What is the difference when there is a group of cells, a group of ants, and a group of individuals?

2) What happens to the systems in the above problem at (a) their moments of sudden death; (b) 10 seconds after their respective sudden deaths; and (c) one year after their respective deaths?

3) Discuss the phenomenological views of philosopher Edmund Hüsserl (1859–1938) as they relate to the philosophical aspects of the entropy.

4) Discuss the views of philosopher Karl Popper regarding the thermodynamics of open systems.

5) Describe the state of our world if there were no entropy.

6) Is entropy per se always undesirable?

7) Make a table with all of the different entropies we discussed in this chapter as rows, and attributes as columns. In the first column enter the defining equation. In following columns place the type of systems, i.e., open/closed/isolated; conservative/dissipative; macroscopic/microscopic/equilibrium/nonequilibrium; linear/nonlinear; other attributes that are meaningful, and cautionary remarks. Indicate which entropies harmonize with one another and which ones appear contradictory.

8) Propose means of measuring each type of entropy discussed in this chapter.

9) In this chapter we discussed a number of different entropies. Would it have been better if each had been given a different name, such as, for example: the Clausius, the Prigogine, the Boltzmann, the Shannon, the Eddington, the Kolmogorov, without ever having mentioned the word "entropy?" Would you have felt more comfortable?

10) Write an essay comparing ergodicity as a branch of mathematics and a condition in physics.

Chapter 9—Dimensions and Scaling

1) It has been suggested that complexity and size are related in both biological organisms as well as in technological devices. How good a guide is scaling in determining the limits of complexity?
2) You are informed that two objects have the following H-B dimensions: 1.33 and 1.77. What, if any, insights do these numerical figures provide regarding the geometry, the structure, and the complexity of the two objects?
3) Evaluate to a first order what the relative H-B dimensions for laminar flow and for turbulent flow might be.
4) Discuss whether the manner in which the continents were formed has anything to do with coastlines having dimensions in the 1.05–1.25 range.
5) The research of Hénon was motivated by astronomical considerations, and that of Lorentz was motivated by the desire to understand the weather. In Table I of this chapter are given a number of dimensions. What do you infer from the fractal dimension of the Hénon map being 1.22 and that of the Lorentz map being 2.06? Discuss whether this means that the Lorentz map is more complex than the Hénon map. Therefore, are you prepared to argue that celestial dynamics is less complex than weather dynamics?
6) Why does the logistic map have such a low dimension?
7) Discuss the verifiability of the Rayleigh–Bénard dimension on the level of complexity of atmospheric phenomena.

Chapter 10—Gallery of Monsters

1) Determine the fractal dimension of a point.
2) Determine the fractal dimension of a straight line.
3) Determine the fractal dimension of a plane surface.
4) Determine the fractal dimension of a cone.
5) Determine the fractal dimension of your hand.
6) Discuss how it is possible that fractal dimensions may exceed 3.
7) Determine the fractal dimension of an equilateral triangle. What will the fractal dimension be if the base is removed so that it looks like a circumflex? Will the fractal dimension change if the equilateral triangle is stood on its apex?
8) Continue the steps of the von Koch snowflake in Chapter 10. How many additional steps did you manage to draw? What shape

would you expect if you were to go to the limit, i.e., n → ∞?
Which increases faster: the perimeter or the area inside it?

9) What are the elements of the array making up the IFS if the triangle is stood on its base, on one of its sides, and on its apex?

10) Determine the fractal dimensions of your initials.

11) Determine the respective IFS arrays for your initials.

12) Develop an alternate form of the Cantor set. Specifically, remove the middle one-fifth instead of the middle one-third. Determine the dimension of this new Cantor set.

13) Develop an alternate von Koch snowflake by adding one-quarter instead of one-third tents. Determine the dimension.

14) What is the fractal dimension of the Peano curve?

15) Explain the structure of the Eiffel Tower in Paris in terms of a three-dimensional Sierpinski triangle.

16) Discuss the fractal dimension of Buckminster Fuller's geodesic structures.

17) Confirm the dimension of the Sierpinsky triangle given in this chapter by performing the appropriate calculations.

18) What are representative fractal dimensions of the class of molecules called "fullerrene's," nicknamed "buckyballs," named after the celebrated Buckminster Fuller?

19) A cylindrical rod having a diameter of 1 cm is 1 m long. What is its fractal dimension? Next, assume that the rod is bent into an L shape so that one leg is 25 cm and the other is 75 cm. What would the fractal dimension become? What would be the fractal dimension if the legs were equal in length, i.e., 50 cm each?

20) Repeat problem 19 for the case of a bar 1 m long and having a cross section 1 cm by 1 cm.

21) Repeat problem 14 for the blade of a metallic roll-up measuring tape. What generalizations follow from Problems 14, 15, and 16?

22) Propose some applications of Cantor dust. Is talcum powder an application of Cantor dust? What about flour or powdered sugar?

23) Is there a conflict in applying the H-B dimension to dynamic systems, considering that time is not apparent in the equation that defines the H-B dimension? Discuss how valid it is to apply the H-B dimension in analyzing time series data? In such situations might the Kolmogorov entropy recommend itself because of its dynamic nature? How about the Rényi entropy?

24) Discuss the following statement: "All strange attractors are fractals, but not all fractals are strange attractors."

25) Must fractals be self-similar?

26) Discuss the applications of chaos, fractals and IFSs to the visual arts, to music, and to literature.

Chapter 11—The Diagnostics and Control of Chaos

1) Is the expression "control of chaos" an oxymoron?

2) Rocket motors may become unstable and thus may be destroyed. Such instabilities are not predictable. Is this a case of chaos? (Hint: The literature on rocket instabilities may be found in publications from the 1950s.) Could the chaos control approach in this chapter be used to dampen the instabilities?

3) Can the Chernobyl and Three Mile Island accidents be ascribed to chaos? Outline the methodology you might pursue to perform the diagnostics. Could the chaos control techniques outlined in this chapter have been useful in averting the problems?

4) Explain why some time series are flat in the normal state, while others exhibit fluctuations. What does the difference tell us about the nature of the respective systems?

5) Imagine that you find yourself in the following position. You are presented with sound data time series taken without interruption since 1980 in Moscow's Red Square. What sorts of observations might you make? For example, at nights there will be fewer fluctuations. What about over weeks, months, and years? What about during the 1991 social upheavals?

6) Why is it so difficult to arrive at useful conclusions by applying chaos theory techniques to the analysis of Dow Jones time series data?

7) During times of election, opinions change. Might the Ising model be used to advantage to make predictions?

8) Using the discussions in this chapter, attempt to design a flexible heart pacemaker.

9) Perform a spectral analysis on the two heart rhythms in this chapter. Determine their dimensions. Is either chaotic?

10) Perform a spectral analysis on the seismograms shown in this chapter. Determine their dimensions. Is there any chaos? Can we attempt to predict anything?

11) Make spectral analyses of the carbon dioxide data given in the two different curves. Also, determine their dimensions. What do the differences suggest? Is any chaos evident?

INDEX OF NAMES

SUBJECT INDEX

$1/f$ scaling, 207
$1/f$ noise, 204–207

A

accuracy, 30
affine transformations, 189
algorithmic theory, 35–37
ambiguity, 128
anatomy of systems and structures, 41–56
arithmetic growth, 82
Arnold's cat, 75
asymptotic stability, 49, 91, 93
atmospheric carbon dioxide, 196–197
attractor, 59
autocatalysis, 78, 98, 103
autonomous system, 44
average rate of information loss, 154

B

Barnsley's chaos game algorithm, 189
barycentric velocity, 49
basin of attraction, 60
Belousov–Zhabotinsky Reaction, 77–79
Bernoulli shift, 74–75
bifurcation, 54
bifurcation map, 117, 210
biological clocks, 79
birth rate, 83
black noise, 206
Bohr's principle of complimentarity, 128
Boltzmann's entropy, 141–144
Boltzmann's H-theorem, 142
Brandenburg concerto, 206
Brillouin–Schrödinger entropy, 144–145
broad band spectra, 203
Brownian motion, 100–101